SUSTAINABLE WATER RESOURCE DEVELOPMENT AND MANAGEMENT

SUSTAINABLE WATER RESOURCE DEVELOPMENT AND MANAGEMENT

A. Zaman, PhD and Md. Hedayetullah, PhD

AAP APPLE ACADEMIC PRESS

First edition published 2022

Apple Academic Press Inc.
1265 Goldenrod Circle, NE,
Palm Bay, FL 32905 USA
4164 Lakeshore Road, Burlington,
ON, L7L 1A4 Canada

CRC Press
6000 Broken Sound Parkway NW,
Suite 300, Boca Raton, FL 33487-2742 USA
2 Park Square, Milton Park,
Abingdon, Oxon, OX14 4RN UK

© 2022 Apple Academic Press, Inc.

Apple Academic Press exclusively co-publishes with CRC Press, an imprint of Taylor & Francis Group, LLC

Library and Archives Canada Cataloguing in Publication

Title: Sustainable water resource development and management / A. Zaman, PhD and Md. Hedayetullah, PhD.

Names: Zaman, A. (Scientist), author. | Hedayetullah, Md., 1982- author.

Description: First edition. | Includes bibliographical references and index.

Identifiers: Canadiana (print) 20210369906 | Canadiana (ebook) 20210369965 | ISBN 9781774630099 (hardcover) | ISBN 9781774639504 (softcover) | ISBN 9781003180494 (electronic bk.) | ISBN 9781000400663 (electronic bk. : PDF) | ISBN 9781000400915 (electronic bk. : EPUB)

Subjects: LCSH: Water-supply, Agricultural. | LCSH: Irrigation. | LCSH: Water resources development. | LCSH: Sustainable agriculture.

Classification: LCC S494.5.W3 Z36 2022 | DDC 631.7—dc23

Library of Congress Cataloging-in-Publication Data

....

CIP data on file with US Library of Congress

....

ISBN: 978-1-77463-009-9 (hbk)
ISBN: 978-1-77463-950-4 (pbk)
ISBN: 978-1-00318-049-4 (ebk)

About the Authors

Prof (Dr.) A. Zaman is an internationally renowned scientist who has been specializing in agricultural water management during his career of 40 years. His lifetime and original research contributions to the field of irrigation water management are on rainfed agriculture, cropping system, watershed management, dryland agriculture, and biodiversity conservation, to name a few. His work is globally appreciated, leading to outstanding teachers in farm universities in India and abroad. He has contributed a large number of original concepts, strategies, techniques that have contributed enormously to the field of agricultural science.

Dr. Zaman served as Irrigation Agronomist and formerly Dean of the Faculty of Agriculture, Director of Research, and Head of the Department of Agronomy at Bidhan Chandra Krishi Viswavidyalaya (BCKV), an agricultural university in West Bengal, India. He is a former Chief Scientist at All India Coordinated Research Project (AICRP) on Water Management (ICAR) and former Associate Director of Research, Regional Research Station, Old Alluvial Zone (presently UBKV, Pundibari, India). He has coordinated and been associated and involved with a postgraduate teaching program at the university and has significant research experience in the field of water management and in academia. He also has considerable administrative experience in different capacities at various time frames.

Dr. Zaman worked as Head of the Irrigation Expatriate Team, Government of Uganda, for a two-year period (1994–1996) under ITEC program being deputed by the Government of India for the design, layout, and installation of farmers' managed small-scale irrigation systems for sustainable crop cultivation. He was also designated as an Indian Farmers Fertiliser Cooperative Limited Chair Professor.

His scholarly publications include more than 150 research papers, most of which were published in peer-reviewed journals having national and international readership. He has also published nine book chapters and seven books with global publishing houses. His academic contribution was profusely appreciated by Dr. Rash B. Ghosh, Charles Townes Prof. of Water Chemistry, with whom he cofounded the International Institute of Bengal Basin and organized international conferences in India and Bangladesh biannually. He organized and chaired many prestigious workshops and chaired many technical sessions at scientific meetings and platforms. He has delivered a number of invited plenary lectures and keynote addresses and is an editorial board member and examiner and question setter for various agricultural universities in India and Bangladesh.

He has 179 publications in proceedings of national and international seminars and conferences held in the country and abroad to his credit. He guided and supervised seven PhD and at least 50 students toward postgraduate thesis work. He also worked as Program Coordinator in the World Bank's West Bengal Accelerated Development of Minor Irrigation Project from 2014–2016, after superannuation from University Service. Dr. Zaman also acted as Emeritus Professor, M S Swaminathan School of Agriculture, CUTM, Paralakhemundi, Odisha, from 2016–2018. Dr. Zaman was also Professor and Founder Incharge, Agriculture of The Neotia University, Kolkata campus, West Bengal from 2018–2020, where he organized agricultural academic course curriculum and introduced an agricultural degree course in the university from the academic session. Presently, Dr. Zaman is working as Director of the School of Agricultural Sciences, Sister Nivedita University, Kolkata, India.

Dr. Md. Hedayetullah, PhD, is Assistant Professor/Scientist and Officer-In-Charge, *All India Coordinated Research Project* (AICRP) on Chickpea, Directorate of Research, Bidhan Chandra Krishi Viswavidyalaya, Kalyani, Nadia, West Bengal, India. He was formerly an agronomist at the National Bank for Agriculture and Rural Development, Balasore, Odisha, India. He was formerly Assistant Professor at the M.S. Swaminathan Institute of Agriculture Science, Centurion University of Technology and Management, Gajapati, Odisha, India. Dr. Hedayetullah was also Assistant Professor at the College of Agriculture, Tripura, Government of Tripura, India.

Dr. Hedayetullah is author and coauthor of 30 research papers, five review papers, 25 book chapters, and six books. He has guided two students for postgraduate degrees and has supervised one research student for postgraduate degree in agronomy. Dr. Hedayetullah has to his credit a number of full-length research papers, popular scientific articles, abstracts, proceedings, and technical bulletins. He has attended more than 30 national and international seminars, symposiums, conferences, workshops, and training courses. He is the principal investigator of ad hoc project titled "Bio-efficacy data generation of diclosulam 84% WDG on groundnut," sponsored by

Deccan Fine Chemicals India Private Ltd. during 2020–2021. He was also principal investigator of ad hoc project titled "Bio-efficacy of PIX 1006 43% WG against major weed of rice," sponsored by PI Industries Pvt. Ltd. during 2019–2020, as well as principal investigator of ad hoc project entitled "Bio-efficacy and phytotoxicity evaluation and residue analysis of paraquat dichloride 24% SL in potato" during 2017–2018 and 2018–2019. He was co-PI of an ad hoc project titled "Creation of seed hubs for increasing indigenous production of pulses in India," sponsored by ICAR-Indian Institute of Pulses Research, implemented at Bidhan Chandra Krishi Viswavidyalaya, Kalyani, Nadia, West Bengal, India, from 2017 to the present. Additionally he was co-PI of an ad hoc project titled "Enhancing pulses production for food and nutritional security improved livelihoods and sustainable agriculture in West Bengal," sponsored by the International Centre for Agricultural Research in the Dry Areas (ICARDA-South Asia and China Regional Programme), implemented at Bidhan Chandra Krishi Viswavidyalaya, Kalyani, Nadia, West Bengal, from 2017 to the present, as well as co-PI of ad hoc project titled "Bioefficacy of herbicide on paddy," sponsored by Indofil and implemented at Bidhan Chandra Krishi Viswavidyalaya, Kalyani, Nadia, West Bengal.

Dr. Hedayetullah is an editorial board member of *Amity Journal of Agribusiness* (AJAB). He is the life member of the Crop and Weed Science Society and the Indian Society of Pulses Research and Development. He is a member of the International Institute Bengal Basin, Berkeley, California, USA. He has gained his first-hand extension experience in different agricultural extension program and farmers-scientists meets, etc., as a resource person. He has appeared on several live TV talks presented at "Krishi Darshan" telecasted by Door Darshan, Kolkata, on irrigation management in crop production. He was the organizer of several front-line demonstration (FLD) farmers' training programs, funded by the Indian Institute of Pulses Research and the Indian Council of Agricultural Research, under All India Coordinated Research Project on Chickpea.

He acquired a Bachelor of Science (Agriculture) degree from H.N.B. Garhwal University, UK, India. He received a Master of Science (Agronomy) from PalliSiksha Bhavana, Institute of Agriculture, Visva Bharati University, Sriniketan, West Bengal, India. He received his PhD (Agronomy) from Bidhan Chandra Krishi Viswavidyalaya, Mohanpur, Nadia, West Bengal, India. He was awarded the Maulana Azad National Fellowship Award from the University Grant Commission, New Delhi, India. He has received several fellowship grants from various funding agencies to carry out his research works during his academic career.

Contents

Abbreviations

BCM	billion cubic meters
BD	bulk density
CCA	cultivable command area
CADP	command area development program
CADS	command area development scheme
CGWB	Central Groundwater Board
CWC	central water commission
dS/m	deci siemens per meter
DVC	Damodor Valley Corporation
ECw	electrical conductivity
ER	effective rainfall
FC	field capacity
FOME	field operation and management efficiency
GCA	gross command area
GW	giga watts
GWR	gross water requirement
IR	irrigation application
IWRM	integrated water resource management
Km3	kilometer cube
MAL	maximum acceptable limit
MCC	maximum capillary capacity
meq/L	milliequivalents per liter
mg/L	milligrams per liter
Mha	million hectare meter
MI	minor irrigation
mmho/cm	millimhos per centimeter
MWHC	maximum water-holding capacity
NWR	net water requirement
NWDPRA	National Watershed Development Project in Rainfed Area
PD	particle density
PMKSY	Pradhan Mantri Krishi Sinchayee Yojana
ppm	parts per million
PWP	permanent witling point
SAR	sodium absorption ratio

TDS	total dissolved solid
WP	water productivity
WR	water requirement
WUE	water use efficiency
µS/cm	micro siemens per centimeter

Preface

India, with a population of over a billion (1.336 as on December 31, 2019), is the world's largest democracy, and traditionally, civilization in India, as around the world, has, for the most part, evolved and developed around water bodies as most human activities and the agriculture and industrial sectors depend on water.

Agriculture is the main source of maintaining the livelihood of 80% of the people of India, and the economy is still primarily based on agriculture. Since agriculture has been used in the process of advancement of society, water has played the most predominant role. The Ministry of Water Resources, Government of India, has predicted the decrease per capita of water availability; it is estimated to decline further to 1465 cubic meters by the year 2025 and 1235 cubic meters by the year 2050. (Per capita gross domestic product [GDP] is a metric that breaks down a country's economic output per person and is calculated by dividing the GDP of a country by its population.)

The challenges of water resources management are severe, because of the ever-rising population, the expected growth in agricultural and industrial demand, the danger of pollution of water bodies, and, over the long run, the impact of climate stress on water resource availability in many regions of India. Water resources management is an important consideration for researchers when addressing pertinent issues such as global climate change, natural disasters, pollution, environmental degradation, and soil health and food security and how to mitigate the occurrence of such situations.

In addition to the procedures to be adopted for rainwater harvesting and its effective utilization, it is essential to explore the possibilities of interbasin transfer of water from surplus areas to deficit basin areas to maximize the full potential of irrigation development.

The following are the important considerations of water resources, its development, and management that need to be taken into account for discussion:

1. Recycling of polluted, contaminated, wastewater, industrial water, saline, sodic, and other problematic water to be used, after proper reclamation, for agricultural purposes for effective utilization in crop cultivation.

2. Agriculture is the main stakeholder of water resource utilization where about 70% of the total water is being used; hence, strategies for efficient utilization of water-saving irrigation technologies for growing crops need to be developed, along with proven technologies for the transfer water to crop growers on a massive scale.
3. The introduction of supplemental and protective irrigation could save at least 30% of irrigation water, minimizing conveyance and other losses, which would aid rainfed areas of the country with higher crop and water productivity.
4. Adoption of microirrigation (drip, sprinkler, bottle, and pitcher) could save 30–60% of water, with a possible 30–50% increased crop yield.

More crops and more returns per drop of water would help to develop more areas under crop cultivation as well as increase crop yield with better water productivity. Multiple uses of irrigation water are the basic purpose for utilizing integrated water resources development and management.

This book, *Sustainable Water Resources Development and Management,* is written for students of agriculture, researchers, policymakers, and teachers engaged in the field of water management. It is organized into 14 chapters. A brief description of each of the chapters is as follows:

Chapter 1 focuses on surface water resources, groundwater resources, water use for crop production, and irrigation water management.

Chapter 2 establishes water resources availability, water demand, and supply and irrigation potential.

Chapter 3 classifies soil water, the impact of forces on water movement, and retention of water in soil.

Chapter 4 looks at issues on water resource development and irrigation command in the country of India along with irrigation potential creation and utilization.

Chapter 5 reviews the irrigation projects in India, particularly the development of major and minor irrigation projects.

Chapter 6 presents the role of water in plant systems and soil water plant relationships.

Chapter 7 addresses water productivity in agriculture and irrigation efficiency. It also focuses on storage efficiency, conveyance efficiency, distribution efficiency, and application efficiency.

Chapter 8 presents the status of surface and groundwater and conjunctive use.

Chapter 9 discusses how irrigation water is measured in different farm fields and farmers' fields to estimate the water requirement of crops.

Chapter 10 provides the definition and conveys the concepts and importance of water quality in agriculture. It also focuses on the impact of water quality on human health.

Chapter 11 presents the objectives and importance of river linking projects in India for the effective solution of diverting "surplus" water to the "deficit" zone.

Chapter 12 discusses water pricing feasibility of users as well as suitable water management technologies adopted and disseminated to farmers for enhancing agricultural productivity in the country with standardization of fixing water prices in agricultural uses.

Chapter 13 discusses wetland management and water productivity.

Chapter 14 discusses water pollution in agriculture and water contamination in urban and rural areas.

The book thus attempts to be a useful document that covered the entire prescribed syllabi of courses on sustainable water resources management.

The authors invite suggestions, modifications, additions, corrections, etc. from readers the improvement of this book toward making it a more authentic and effective resource to fulfill the national aspiration of efficient sustainable water resources management.

—Prof (Dr.) A. Zaman
Dr. Md. Hedayetullah
Dated: 18 September, 2020

CHAPTER 1

Water Resource Development

ABSTRACT

Water is a vital need for sustaining the soil health and supporting life, and the agricultural sector is a major stakeholder of water resources. Despite the fact that over two-third of the earth's surface is spread with water, currently about 450 million individuals in 29 countries face water shortage and at least 20% of additional water would be required to satisfy the entire water demand for extra 3 billion population by 2025. Because of the fact that among the entire available water, about 97.5% of water is not usable, because of its salinity from the ocean and 2.49% is ice, and only 0.01% is technologically offered and economically accessible water either in different surfaces or in well water sources for human use as well as water in watercourse, lake, and water bodies. Water is running out because of indiscriminate use, especially in agriculture and other uses. Water resources stand one in all the main crises and a threat of dire consequences ensuing from economic condition, hunger, eco-system destruction, soil health condition changes fastening along different severe world issues. Hence, the agricultural water management should be adequately addressed with integrated approach where conservation and transfer of water with higher economic values be obtained along with environmental advantages. Agriculture under rainfed condition is also associated with primarily conservational land treatments to visualize erosion and land degradation must also to be considered as integral part of water resource development and management.

1.1   INTRODUCTION

India, with a population of over a billion, has the world's largest democracy; traditionally, civilization in India, for the most part, is established around water bodies, as most human activities in agriculture and industries depend upon water . The water scenario in India appears to be going from bad to worse. Not only is there a growing scarcity of water within the country but the agriculturally vital states like Punjab, Haryana, Tamil Nadu, and Rajasthan also face a fall in groundwater levels. Although the per capita convenience of available water in India in 1951 was 3500 m³, in 1999, it fell to 1250 m³. This is in line with the Ministry of Water Resources, which predicts a decrease to 662 m³ per person in 2050. The challenges of water resources management facing severity with pertinent question of the chance that area unit intensified over time because of ever-rising population, expected growth in agricultural and industrial demand, the danger of pollution of water bodies, and over the long run, the impact of climate stress on water resource availability in many regions of the country. Historically, efforts to deal with water supply problems have centered around major and medium irrigation and come as attainable measures to mitigate the issues. However, the use of water is characterized by increasing the dependence on groundwater for irrigation in agriculture. The annual extraction is about 7416 billion cubic feet of groundwater in India is by far the maximum in the world. These days, groundwater provides water source for over 70% of the net irrigated area. It accounted for over 85% of the addition to the irrigated space within the last 30 years. The area irrigated by canals and tanks has actually undergone a decline even in absolute terms since the 1990s in a rustic of such immense physiographic, hydrogeologic, and demographic diversity, and also different levels of economic development, water balances for the country as a full area unit of limited value since they hide the existence of areas of acute water shortages and additionally issues of quality. The major contributor to the current speedy depletion in the water level is the overwhelming dependence on deep drilling of groundwater through tube wells, which these days account for over 40% of irrigation. Indeed, we have a tendency to area unit about to coming into a vicious infinite regress situation, wherever a trial to resolve a problem solution. The foremost placing example of this in India is that the accrued reliance on tube wells each for irrigation and drinking, not recognizing that one will probably jeopardize the other. This leads to the development of villages "slipping" back when being covered by rural drinking water program.

India occupies only 3.29 million km^2 geographical area that forms 2.4% of the world's surface area; it supports over 15% of the world's population. The population of India as on March 2016 stood at 1332 billion. Thus, Asian country supports 1/6th of the world population, 1/50th of the world's land, and 1/25th of world's water resources. India additionally includes a livestock population of 500 million that is around 20% of the world's total livestock population. Over half of these livestock are cattle, forming the back bone of Indian agriculture. The entire utile water resources of the country are assessed as 1086 km^3.

1.1.1   SURFACE WATER RESOURCES

In the past, many organizations and people have calculated the water availability in India. Recently, the National Commission for Integrated Water Resources Development calculated the basin-wise average annual flow in Indian watercourse systems as 1953 km^3. Usable water resource means the quantum of water ready for withdrawal from its place of natural prevalence. Inside the restrictions of physiographic conditions and sociopolitical surroundings, legal and constitutional constraints, and therefore, with the technology of development on the market at this time, the usable amount of water from the surface flow has been assessed by numerous authorities otherwise. The usable annual surface water of the country is 690 km^3. There is respectable scope for increasing the employment of water within the Ganga–Brahmaputra basins through the construction of storages at appropriate locations in the neighboring countries.

1.1.2   GROUNDWATER RESOURCES

The annual potential natural groundwater recharge from downfall in India is around 342.43 km^3, which is 8.56% of the total annual downfall of the country. The annual potential groundwater recharge augmentation from canal irrigation system is about 89.46 km^3. Thus, the total replenishable groundwater resource of the country is assessed as 431.89%. After allotting 15% of this quantity of water for drinking purpose, and half-dozen km^3 for industrial functions, the remaining will be utilized for irrigation functions. Thus, the available groundwater resource for irrigation is 361 km^3, of which the usable quantity (90%) is 325 km^3. The estimates were done by the Central Groundwater Board (CGWB) of the total replenishable groundwater resource which provides provision for domestic, industrial, and irrigation uses and utile groundwater resources for future use. The basin-wise per capita water availability varies between 13,393 m³/year for the Brahmaputra–Barak basin to about 300 m³/year for the Sabarmati basin.

1.2   WATER RESOURCES

More than two-third of the earth's surface is roofed with water, and currently, close to 450 million individuals in 29 countries face severe water shortage and a minimum of 20% of additional water would be needed to feed further 3 billion populations by 2025. About 97.5% of ocean water is not usable because of its salinity and 2.49% is fastened up in ice and only 0.01% is technologically offered and economically accessible either from surface or well water sources for human consumption. Thus, water-saving technologies developed to this point are incredibly pertinent to propagate among the users. Water is a vital natural resource for sustaining the surroundings and supporting life wherever the agriculture is that the major user of water resource. Water is an imperative for long-term agricultural growth and development. Previously, rainfed agriculture was associated primarily with conservational land treatments to visualize erosion and land degradation. But within the areas of medium to sufficient downfall, there's ample scope of conserving excess rain water through appropriate water storing structures made for this purpose for its subsequent uses as irrigation or to use life-saving irrigation to the crops that ought to conjointly return below the horizon of water management for crop production. Though there's a distinction within the objectives of water management wherever there's plethoric availability of water from the resource that has continuous flow connected with perennial sources or lakes and rivers which provide restricted or limited supply through the assortment of excess rain through water storing structures. The former is related to get the maximum amount as yield of crop per unit area under water application and the latter is related to increasing the productivity of crop per unit of water application even under rainwater reuse and its effective management for the aim of irrigation. Hence, rainwater and irrigation water are thought of to be equally vital in the context of agricultural development of the country to induce higher water use potency and correct utilization of water resources (Zaman et al., 2016).

Water is a vital need for sustaining the soil health and supporting life, and agricultural sector is the major consumer of the water resource. Despite the undeniable fact that over two-third of the earth's surface is covered with water, currently, about 450 million individuals in 29 countries face water shortage and a minimum of 20% additional water would be needed to satisfy the entire water demand for the extra 3 billion populations by 2025. Of the entire available water, about 97.5% of it is not utilizable because of the salinity of ocean-sea and 2.49% is fixed up in ice and 0.01% is technologically offered and economically accessible water either in different surface or well water sources for human uses as well as water in watercourse, lake and water bodies (Table 1.1). The water is running out because of its indiscriminate uses, water resource stands one in all the main crises and a threat of dire consequences ensuing from the economic condition, hunger, ecosystem destruction, soil health condition changes fastening along different severe world issues. The world population might be water stressed by 2025. With this view, conservation, development, and management of water for agricultural and industrial sectors possess first and foremost importance. The agricultural water management should be integrated with adequate address to conservation and transfer of water to uses with higher economic obtain similarly as environmental advantages.

TABLE 1.1   World Water Availability Pattern.

Forms of water	% availability
Sea and ocean	97.208
Ice and iceland	2.150
Atmospheric water	0.006
Groundwater (under the ground)	0.625
Surface water (lake, river, pond)	0.001

Rainfall is the main supply of water that varied in different agro-ecological zones. The quantity of annual downfall exhibits certain typical characteristics of distribution of the rainfall. India received about 80% of the entire annual downfall at intervals of 3-month time and the average annual downfall comes about 1170 mm compared with the 800 mm world average. The entire rainfall hours is calculated as 200 h whereby over 500th occurred at intervals of 20–30 h time. The wastage of water in the main part through runoff became ineluctable. The infiltrated part of downfall sometimes turns into soil water that is used by the vegetations, some parts of it are evaporated and the rest has gone down as groundwater or appeared as springs and streams whereby the main part had flown as runoff on the land surface. Nearly, 20% of the runoff water might be attainable to utilize effectively. As a consequence, over one-third of the total area became drought prone in our country.

1.3   STATUS OF GROUNDWATER

It is a fact that nature's means of storing water is underground, wherever about 98% of the whole world's water is in liquid form and also as fresh water. And about 2% of water is stored in streams and lakes, which regularly are fed by groundwater as a dependable supply, less affected by the vagaries of climate. Indiscriminate use of groundwater significantly with the introduction of high irrigation responsive photoinsensitive short period rice varieties throughout the year caused a drastic uplift of the groundwater level and posed many sophisticated issues. Groundwater reservoir was primarily replenished by the annual precipitation received in a specific area. The rate of water flow in to the groundwater reservoir was relied on the pattern of rain, runoff, stream flow, porosity of the soils, and earth materials prior to the reach of water up to the zone of saturation. The overall water resource of the country comes to about 400 million hectare meter (m ha m) and of which the total quantity of surface flow throughout the year in our country is calculated to be about 180 m ha m, out of which about 105 m ha m comes from the annual precipitation (Table 1.2). The calculated water recharge per annum is about 67 m ha m, out of which about 50 m ha m comes from the rainfall. There is enough scope to extend the quantity of rechargeable water adopting ways. Introduction of high-yielding irrigation-responsive sorts of rice and intensive cropping system, three to four crops in a year within the same piece of land needed large quantity of water. The water was provided from undergroundwater resources. The deep and shallow tube wells were dug indiscriminately. As a result, the depletion of groundwater level is severe.

In this view, it had been necessary to access the maximum amount because the excess water may increase groundwater table during high to medium rainfall; the chances of incidence of flood would be less. The groundwater level would come back to rise, the recycling of water throughout the period of drought would become easier and safer. There was clear indication that many conservation and water-harvesting measures might increase the groundwater recharge up to an explicit limit. In India, groundwater recharge from open wells and pits are quite common practices in Odisha, Gujarat, and Rajasthan. As a result, larger exploitation of water before *kharif* season provided scope of larger amount to infiltrate throughout the rainy season. The utilization of surface water would become additionally to be more practical and useful to mitigate the arsenic-like issues. Because the risk of contamination of arsenic is terribly less in surface water, the flowing surface water is the safest supply that is free from arsenic. In this context, it is essential to store maximum water underground for its purification and starting off water at would like while not hampering the groundwater level depletion.

TABLE 1.2 Water Resources Scenario in India.

Annual average precipitation	1140 mm
Total available water	400 million ha
Net area sown	145 million ha
Gross cropped area	175 million ha
Irrigated area	70 million ha
Water demand for irrigation	46 million ha

India is the only country in the world having abounding water resources as it is evident from the demand and provide water resources within the country. Water resources demand is lesser than that being provided. However, water scarcity became evident from frequent incidence of drought at completely different locations of the country primarily because of the lack of correct conservation and maintenance, moreover, as uneven distribution of annual rain. Because it was not attainable to collect excess rain throughout heavy to medium peak rain amount, either that would have caused the disasters in variety of flood or flown to the ocean and ocean or followed by drought within the amount once there is no risk of the occurrence of rain. The erratic and uneven distribution of annual rain including lack of correct water harvest technologies, besides improper management of water resources at completely different locations of the country is chargeable for such things. The integrated water resources development and management that ought to have some specific tips to the farmers, which can offer with clear under-standing of putting in irrigation system, supported their own resources in accordance with the actual things of their cultivable land besides economic and even-handed use of irrigation with adoption of water-saving technology for sustainable crop production in agriculture.

1.4 CROP WATER REQUIREMENT

The demand for water to crop usually comes from evaporation and transpira-tion, which is commonly known as consumptive use, and is associated with unavoidable conveyance losses during application and some special needs during land preparation and transplanting. The supply of water to meet the crop needs usually comes from effective rainfall, irrigation water application, and contribution from storage moisture at the soil profile.

Equationally, WR = IR + ER + S where WR = Water Requirement, IR= Irrigation Application, ER= Effective Rainfall , and S = Storage moisture in soil profile.

So, IR = WR − (ER + S).

To calculate the water requirement of the crop, it is essential to know the existing moisture status of the soil at sowing and harvesting on which the crop is growing. It needs the knowledge of root system and rooting habit of the particular crop and procedure to determine the moisture content in the different soil profile up to root zone depth of the crop. The crop usually extracts moisture for essential activities through the root system. Water comes up to the tip of the leaves to complete the food production processes and goes down to the root again.

So, before going to apply irrigation, it is very important to know:

- The crop which the farmer wants to grow with its rooting habits.
- The physiological critical growth stages.
- The existing moisture status of the soil profiles up to root zone depth.

1.5   WATER BALANCE

It is very important to harvest excess rain water which runs as runoff during medium to heavy rainfall period for its utilization in crop production during rainless period and development of rainfed agriculture through scientific management of rain water, particularly in the East African countries. The gains of water would come from total rainfall over a period of application of water as irrigation. The losses constitute a total runoff (R), downward draining (D), Evaporation (E), and Transpiration (T). Thus, the balance sheet shows as Surplus and Deficit over a particular period and it uses to give a changed situation of stored moisture in the soil profiles as well as in the plants.

Equationally,

Gains = P (rainfall) + I (Irrigation); Losses = Runoff (R) + Downward Drain (D) + Evaporation (E) + Transpiration (T); Gains − Losses = S + V, where S and V are changed moisture status in soil profiles and plants, respectively.

1.6   WATER USE FOR CROP PRODUCTION

To get a proper understanding on water use, it needs experimentation and observation on the basis of scientific information on these aspects mentioned above. Only the research findings on water requirement of crops and irrigation water use for crop production would lead to formulate a guidelines on effective utilization of irrigation potential, whatsoever, it has been created. The economic and judicious water application was also ensured by these guidelines. So, the entire principles of irrigation, its theories and practices, depends on:

1. How much water to apply? i.e., Measurement of irrigation water.
2. When to apply the supplemental water as irrigation?, i.e., proper timings and scheduling of irrigation application
3. How best to apply irrigation? i.e., methods of water application to the crop for sustainable production.

1.7   IRRIGATION WATER MEASUREMENT

To establish the water requirement of crops, excluding it from the rainfall occurred and moisture present in the soil profile, the requirement of water is determined. The total amount of water to be supplied as irrigation is usually divided with time of intervals throughout the growth period, which indicates the amount of irrigation to be applied at each irrigation. It mainly depends on the rooting habits of the crop and moisture holding capacity of the soil, so it is very important to determine the moisture content of the soil profile and also it is very pertinent to know the moisture-holding capacity of the soils.

Soil sample placed in dry air oven at 105–110°C temperature for 16–18 h up to a constant weight, it gives the moisture content, multiply it with soil depth from where the soil sample was collected (cm) and apparent specific gravity (bulk density, g/cm^3), the moisture content in the soil profile can be determined. This value is subjected to change with tillage operation, cropping pattern, and manure application. So, it is advisable to check on the bulk density values of the soil, time to time, before calculation of the moisture content in the soil profile.

The following are the preconditions of irrigation water measurement or the amount of water to be applied at each irrigation:

•   Occurrence of rainfall, if any
•   Calculation of soil moisture and bulk density

- Rooting habit of the crop to know the effective root zone depth
- Moisture holding capacity of the soil at each soil depth.

The moisture holding capacity is commonly known as field capacity and subtracting it from the moisture content of the soil profiles, the amount of water can be measured directly by water meter. There are several indirect methods of measuring irrigation water like orifice, v notch, and Parshall flume. These are very simple devices to measure irrigation water to crop field during its application, which needs only calibration with time factor.

1.8   IRRIGATION SCHEDULING

It is equally important to fix up the time intervals of water application to crop field. Several criteria are usually deployed to find out the proper timings of water application. The soil tensiometer, resistant or gypsum block, and neutron moisture meter are used to give information on moisture status at particular root zone depth of soil profile which indicated the moisture availability in any specific zone. These equipments as well as moisture depletion method could be employed to make irrigation schedule of a crop. The cumulative daily evaporation value from any standard open pan evaporimeter also may be used as indices to irrigate the crop at proper timings. But when the water supply is limited, application of water at identified critical growth stages may be performed to make the irrigation schedule to the crop.

1.9   WATER APPLICATION METHODS

Selection of appropriate method of irrigation water application to the crop field depends on soil type (clay, silt, sandy loam, loam, etc.), topography (undulating or flat), availability of water, and crop management practices adopted. The appropriate method would lead to give good storage and distribution efficiency of water.

Surface, subsurface, sprinkler, and drip (trickle) are commonly classified methods where the surface method includes border strip, check basin, and furrow according to field layout and availability of water through gravity flow. Surface methods are usually employed to comparatively flat or leveled land with gentle slope having advantages of gravity flow by selecting suitable water lifting devices or making diversion from the continuous flow of water from perennial sources.

Sub-surface methods are not very common in practices due to its difficulties in lying out and required construction work. Sprinkler (overhead) method is the application of water as similar to rainfall with rotary heads or perforated pipes under the pressure created to the nozzles, holds good to any bunch crops like cereals (other than rice) and oilseeds. The drip (trickle) methods involve carrying water through small diameter plastic tubes with drippers, which emit water drop by drop near base of the plants/roots irrespective of topography and holds good and much economic to the horticultural crops like mango, guava, citrus even for coffee, and banana. Though initial installation expenditure are more to the sprinkler and drip system, it would be more economic in the long run by increasing yield with much water saving, less power required, much saving in laborers and annual maintenance having effectiveness under any topographical situation at different landscape.

1.10   WATER CONVEYANCE AND IRRIGATION EFFICIENCY

Water from the sources, when it reaches to the crop field as irrigation, it requires any distributory system like watercourses or channels. It involves a loss of water through the distribution system associated with percolation, evaporation, and losses occurred by runoff at borders and furrows. So, the ratio of actual water that has been utilized as irrigation out of total water delivered from the source is known as irrigation efficiency. The ratio between the water stored in the root zone depth and the total water reached to the crop field is known as water application efficiency. The ratio between water stored in the root zone depth and the actual requirement of water prior to irrigation of water application is known as water storage efficiency. These could be transferred and termed as operational or economic efficiency of the system.

The annual global flow of water ranges from 4000 to 4700 m ha m, of which 1400 m ha is utilizable for irrigation and other purpose. The present level of utilization of available potential is just around 25%, of which irrigation constitutes about three-forth. As per estimates of the National Commission on Agriculture, the average annual rainfall over the total geographical area of India is around 370 m ha m, added another 30 m ha m, which is brought in from catchment area lying outside the country that makes 400 m ha m. The amount is estimated that about 85 m ha m of water soaks into the soil (contributing to the profile storage/groundwater) and 130 m ha m evaporates into the atmosphere. The remaining 180 m ha m constitutes the average annual flow in different rivers, of which only about 70 m ha m is utilizable owing to the topographical, hydro-logical, climatic, soil, and other limitations. If policy decision is taken on groundwater development in command areas, an additional area of about 1 (one) million hectares in Andhra Pradesh, Tamil Nadu, West Bengal, Odisha, Bihar, and Assam could be brought under irrigated agriculture. In *boro*/summer, even crops like groundnut, maize, potato, and linseed could successfully be grown under rainfed lowland ecology after the rainy season rice on the availability of groundwater as well as residual mois-ture. The conjunctive use of surface and groundwater as well as residual moisture including water productivity assessment under rainfed ecology should come under the purview of integrated water resource management for sustainable agriculture.

Agriculture accounted for at last 90% of annual withdrawal of renew-able water resource. The factors involving exploitation of natural resource, irrigation are by and far the most important. Out of total 253 m ha irrigated land in the world, China and India alone have over 100 m ha. In pursuit of water demand, the competitive supply of irrigation water is declined rapidly due to lack of sufficient budget allocation for infrastructure development and degradation along with lack of proper maintenance of existing infra-structure for irrigation. The overexploitation of groundwater resource is also playing a crucial role as the indiscriminate use of irrigation water from groundwater sources, resulted in continuous dropping of water table with possible resultant health and environment hazards effect. The per capita withdrawal of water resource of about 612 m^3, is too high as compared with only 298 m^3 in South Korea and 462 m^3 in China, where India, South Korea, and China are having renewable water resource availability which is 2464, 1452, and 2427 m^3 per year person, respectively. The per capita availability of water resource declined by 40–60% in most of the Asian countries over 1955–1990 periods.

Out of 400 m ha m annual rainfall in our country, only 50% is exploitable for various purposes in agriculture, industry, navigation, forestry, and power generation wherein about half of water resources are used for irrigation. This leads to an ultimate irrigation potential projected up to 114 m ha, of which only 92 m ha has already been achieved. In Eastern India, many multipurpose irrigation projects were developed due to presence of a number of rivers and due to high rainfall intensity (Table 1.3). The groundwater resources are also rich still now that have significantly contributed toward the development of irrigated agriculture.

TABLE 1.3 Source-wise and State-wise Net Irrigated Areas in Eastern India.

State	Source (in thousand ha)			
	Canals	Tanks	Tube wells	Other source
Assam	362	NA	NA	210
Bihar	1072	117	1794	641
Odisha	949	305	836	NA
West Bengal	717	263	712	219
NEH	28	5	4	240

In Eastern India, only 36% of the total net cropped area (7% in Mizoram and 49% in Bihar) has been brought under irrigation exploiting all possible sources. In West Bengal, there are three major irrigation projects operating. These are Damodar Valley (Bankura, Hooghly, Burdwan, and Howrah), Mayurakshi (Birbhum, Murshidabad, Burdwan), and Kangsabati (Bankura, Midnapore, and Purulia). A part from these, Teesta-Torsa and Subarnarekha Barrage projects are also started and under construction. Work on 18 medium irrigation schemes is continuing in the district of Birbhum, Bankura, Purulia, and Burdwan. In addition, water-harvesting structures, de-siltation basins, runoff retention tanks, and other measures are being taken into serious consideration. National Watershed Development Project in Rainfed Area (NWDPRA) involved the common farmers in conservation of land and water wherein at least 95,000 ha of land have been brought under the program. Thus, the irrigation potential created up to 45% or arable land to cover in the state. Though the potential created at the cost of huge exchequers are being underutilized. Indiscriminate use of groundwater particularly with the introduction of high irrigation responsive photoinsensitive short duration rice varieties growing throughout the year caused a drastic depletion of the groundwater table which results to have complicated problems. Groundwater reservoir was primarily replenished by the annual precipitation received in a particular area. The rate of entering water to groundwater reservoir depended on the pattern of rainfall, runoff, stream flow, permeability of the soil, and earth materials present prior to reach the water up to the zone of saturation. Total amount of surface flow throughout the year in our country is estimated as about 180 m ha m, out of which about 105 m ha m comes from annual rainfall. An estimate of groundwater recharge per year is about 67 m ha m, out of which about 50 m ha m comes from rainfall. So, there was enough scope to increase the amount rechargeable groundwater by adopting ways and means. Introduction of high yielding irrigation-responsive varieties of rice and intensive cropping system, three to four crops in a year in the same piece of land required huge amount of water. The water was supplied from groundwater sources. The deep and shallow tube wells were dug indiscriminately. As a result, not only the depletion of water table occurred to be severed, but also per capita availability of water would come down from 2100 m³ at present to 1700 m³ in 2030 (Table 1.4).

TABLE 1.4 Water Availability and Water Use Indicators (Asian Countries).

Country	Renewable water resource (m³ per year per person)	Per capita withdrawal (m³)	Withdrawal as % of availability	% share of agriculture
China	2427	462	19.1	87
India	2464	612	24.8	93
Indonesia	14,020	465	3.3	94
Bangladesh	11,740	211	1.8	96
Thailand	3274	599	1803	90
Vietnam	5638	81	1.4	78
Myanmar	25,960	103	4.0	90
Philippines	5180	693	13.4	61
South Korea	1452	298	20.5	75
Pakistan	3962	1250	31.5	98
Nepal	8686	155	1.8	95
Malaysia	26,300	765	2.9	47
Sri Lanka	2498	503	20.1	96
Cambodia	10,680	69	0.6	94

1.11 ON-FARM WATER MANAGEMENT

Adoption of suitable and economical water management technologies are essential to maximize crop production and minimizing gap between irrigation potential created and used. Application losses due to conveyance, methods and scheduling of irrigation also are effective in this direction. Selection of crop(s) and crop sequence(s) also played an important role for the enhancement of water use efficiency through crop cultivation and related agronomic practices. Water harvesting, storing, distribution, and utilization of irrigation water are the basic parameters of on-farm water management. Optimum scheduling of irrigation, suitable method adoption, conjunctive use of rain, surface and groundwater for crop cultivation, having improved agro-technology adoption, and provision of drainage are essential. Application of proper amount of water at proper time increased the water use efficiency and crop yield maximization with given amount of water reducing evaporation and deep percolation. Scheduling of irrigation with limited water availability is a big challenge to the irrigation experts that needs rigorous research. It is considered that the irrigation system is a complex sociotechnical phenomenon that involved interaction between a physical, environmental, and technological in the process of adoption. However, the participatory action research appeared to be more effective and communicative measures are in the developmental processes. It is described that the participatory action research is a method of merging both developmental intervention and research activities as an effort to involve the users in systemic process of change to make the sociotechnical system more effective and efficient. Majority of the farmers are reluctant to apply irrigation scheduling technologies as they do not expect a net benefit out of its application. Some workers elaborated their work on-farm participation and encouraged the farmers to adopt irrigation scheduling as cost-benefit approach. They formulated user-friendly format for a wide variety of crops irrigated in South Africa. The on-farm irrigation management and irrigation service are improved considerably by involving farmers in water management practices and introducing an acceptable water-saving irrigation schedule for growing crops and increasing agricultural production. On-farm water management with emphasis on active participation of the farmers at each and every stage is the corner stone of success.

A series of on-farm water management research activities was conducted as a collaborative project of Indian Council of Agricultural Research, New Delhi and Ministry of Water Resources, Government of India in some of the selected command area either by package intervention or conducting component technology field trials in the irrigation command. On-farm participatory field trials brought positive improvements toward better water use efficiency following the recommended methods of water management practices in the farmers' field in Damodor Valley Corporation (DVC) command. Several research workers also reported success story of efficient soil and water management for enhancing agricultural productivity in diverse soil groups of India.

1.12 CONCLUSION

Water resources management is an important consideration to the researchers to address the issues like global climate changes, natural disasters, pollution, environmental degradation, and soil health and food security to fit the system in mitigating the occurrence of such situations. In addition to the procedures to be adopted for rainwater harvesting and its effective utilization, it is essential to explore the possibilities of inter basin transfer of water from surplus to deficit basin to increase maximum water resources to get the full potential of irrigation development.

1. It is also essential to find out the ways and means to use the contaminated as well as reclaimed water for its effective utilization in crop cultivation.
2. Efficient utilization of water-saving irrigation technologies for growing crops.
3. Introduction of supplemental and protective irrigation that could save at least 30% of irrigation water minimizing conveyance and other losses.
4. Adoption of microirrigation could save 30–60% of water with 30–50% increased crop yield. Save water and increase the area under crop cultivation as well as more crop yield with less water application is the basic purpose of adapting integrated water resources management.

KEYWORDS

- **water resources**
- **surface water**
- **groundwater**
- **water balance**
- **crop water requirement**

CHAPTER 2

Irrigation Water Resources

ABSTRACT

In India, the average annual rainfall is 1194 mm, but the intensity of rainfall varies from region to region. The geographical area of India is 328 million hectares, and annual rainfall amounts to 392 million hectare meters. This may round off to 400 million hectare meters (m. ha. m) by including the contribution of snowfall. Out of 400 m. ha. m of rainfall, 75% is received during southwest monsoon period (June to September). A major part of water infiltrate becomes runoff and leach into the soil, while 70 million hectare meters is lost as an evapotranspiration. Water is indispensable commodity for human, animals and all plant life in the universe and being essential part of protoplasm; it is an important ingredient of photosynthesis in food manufacturing within. Water is also required for translocation of nutrient and dissipation of heat in different media. The growing demand for water due to depletion of the available water, assured supply of good quality water is becoming a growing concern. The main objective of the water policy is to optimize its availability of water for different purposes, especially for the supply of water for drinking (domestic), food production (agriculture), livestock (bovine), as well as for power generation, navigation, and various commercial, industrial and domestic uses. Equal importance should be given to increase water use efficiency, equity, and sustainability of water productivity through worldwide. Sustainability issues are to be highlighted in the light of the declining per capita water availability, as well as the pollution of water through human intervention in different water sectors. Strategy and action plan is required for harnessing the available water resources while also protecting and conserving the valuable natural resources in befitting manner.

2.1   INTRODUCTION

About 97% of world's total water is not useful for the irrigation purposes. Of the total quantity of world's water, about 2.5% is available as sweet or fresh water; those are present in the ice cap and glaciers, whereas very little percentage is present in the earth surface, river, and environment. This fresh water can be utilized for irrigation purpose in agriculture sector. The major sources of the total sweet or fresh water present for irrigation purposes or for human domestic consumption are rainfall and snowfall. Rain water passes through river and stream and some portions of it store as in the form of ditch, depression, pond, and water-logged area. Some portion of rain water goes down in the form of infiltration and store in different water-bearing strata. All these stored sources of fresh water are used for irrigation and human domestic purposes. Rain water collected by means of natural or artificial process is diverted to main agriculture field with the help of submersible or pump set. Rainfall is stored as a groundwater through recharging during rainy season. On an average, annual rainfall is received about 400 million hectare meter (m ha m) and get recharged about 215 m ha m through infiltration process below the soil surface. Out of 215 m ha m, about 165 m ha m get adsorbed, which is essential for the growth of vegetation grown on earth naturally. After receiving the rain water or irrigation water, soil adsorbed to field capacity. Later on, it gets percolated into the soil to water table and raises groundwater table.

The water is a precious commodity, finite, and in view of growing demands day by day, ultimately scarce natural endowment. The main objective of the water sector policy is to optimize its availability for different purposes, especially for the supply of water for drinking, food production, livestock, as well as for power generation, navigation, and various commercial, industrial purposes, and domestic uses. The equal importance is given to increase water use efficiency, equity, and sustainability in the use of water through worldwide. Sustainability issues are particularly important in the light of the declining per capita availability of water, and the pollution of water through human intervention in different water sector. Per capita water resources are not static, in view of this precious commodity and its utilization should be proper manner. India, which has 2.45% of land resources, has roughly 4% of the world's fresh water resources, whereas the country's population is 16% of the world's population is big challenges in near future. Most of the rainfall (76%) is received during June to September, which is south west monsoon rain and except in the state of Tamil Nadu which falls under the influence of the northeast monsoon during winter October to November months. More than 50% of precipitation takes place in about 15 days and less than 100 h altogether in a year which reflected on society as flood. India receives on an average annual rain of about 4000 km³ Billion Cubic Meters (BCM) including snow fall or ice fall. The seasonal rain receives about 3000 BCM which is about 75% of average rain fall. Out of this, the average annual water flow available in rivers is around 1953 BCM. Many other rains also occur throughout the year and it depends upon the geographical location.

2.2 WATER RESOURCE AVAILABILITY

The wide diversity exists of inland water resources of India due to several reasons. These areas are mainly distributed over 12 states as given in Table 2.1. The Table 2.1 shows the state-wise inland water resources. It is known that India has rivers and canals of total 15,210 km length which play a significant role in Indian irrigation system. The total water bodies about 29.07 lakh ha reservoirs are available across the country, among these 24.14 lakh ha water bodies are covered by tanks and ponds for different purposes. It is also observed that 7.98 lakh ha water is available through flood plain lakes and 12.4 lakh ha water is available through brackish water in the country. Mostly, water bodies are available through reservoirs which is the major concern in irrigation sector. When we considered the state-wise availability of water,

TABLE 2.1 Water Resource Diversity.

State	River and canal (km)	Water bodies (Lakh ha)				
		Reservoirs	Tanks and ponds	Flood plain lakes water	Brackish water	Total
Odisha	4500	2.56	1.14	1.8	1.8	9.80
Andhra Pradesh	11,514	2.34	5.17	–	0.6	8.11
Karnataka	9000	4.4	2.9	–	0.1	7.40
TN	7420	5.7	0.56	0.07	0.6	6.93
WB	2526	0.17	2.76	0.42	2.1	5.45
Kerala	3092	0.3	0.3	2.43	2.4	5.43
UP	28,500	1.38	1.61	1.33	–	4.32
Gujarat	3865	2.43	0.71	0.12	1	4.26
Maharashtra	16,000	2.79	0.59	–	0.1	3.48
Arunachal Pradesh	2000	–	2.76	0.42	–	3.18
Rajasthan	5290	1.2	1.8	–	–	3.00
MP	17,088	2.27	0.6	–	–	2.87
Others and UTs	84,415	3.53	3.24	1.39	1.2	9.36
Total	195,210	29.07	24.14	7.98	12.4	73.59

Source: Department of animal husbandry, dairying and fisheries.

then it is minutely reflected that Odisha and Andhra Pradesh have the largest water bodies than other states in India.

2.3 WATER RESOURCES AND IRRIGATION DEVELOPMENT IN INDIA

Water is indispensable commodity for human, animals, and all plant life in the Earth. Most of all the life needs water and almost 85% of water is present in the system, among these, some of them contain more than 90% of water. Water is the essential part of protoplasm and it is an important ingredient in photosynthesis. Key rule for dry matter accumulation of plants is 400–500 L of water is required for the accumulation of 1 kg of plant dry matter. Water is needed for translocation of nutrient within the plant system. The average annual rainfall of India is 1194 mm but the intensity of rainfall varied from region to regions. The geographical area of India is 328 m ha, and annual rainfall amounts to 392 m ha m. A total of 400 m ha m fresh water is available in the earth by different sources for domestic, industrial, and agricultural use. Out of 400 m ha m of rainfall, major portion comes from the southwest monsoon period as depicted in Figure 2.1. A major part of water infiltrates, runoff, and leach into the soil, while 70 m ha m is lost as an evapotranspiration.

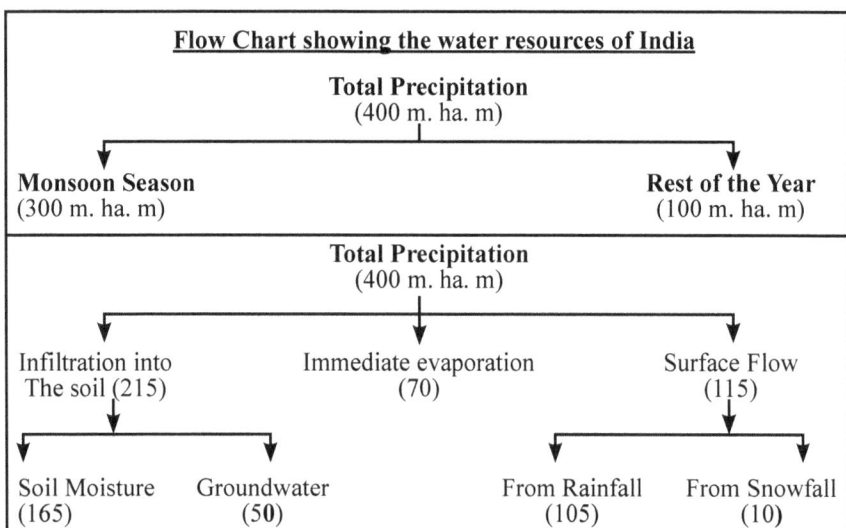

FIGURE 2.1   Water resources of India.

2.4   WATER DEMAND AND SUPPLY

India uses more water than any other country in different aims. Indians are the largest freshwater users in the world which is readily available in many forms in all the year round. In India, major portion of water is used from groundwater which plays an important role in the development of industry and agriculture and results in economic growth. Other sources of water, that is, river, canal, watershed, lake, pond, and farm pond also supply for the nation for different purposes. Gradually, water and air pollution is increasing; good quality of water demand is so high in different sector. The supply of good quality of water is low or has limited resource. Continuous water supply is essential for the survival of the nation. The water consumption in different segment or sector of India has been assessed by the National Commission on Integrated Water Resources Development during 2010. As per the national commission, major portion of demand in India is for irrigation where agriculture sector is the main stakeholder of water resource utilization, that is, drinking water, and industrial uses etc. Water requirement of the country is projected to very soon overtake the availability of water as population is increasing and uses of domestic as well as agricultural uses are much more. Many parts of the India are facing water crises; it has seen deficiency of water and sufficiency too, but excess water about 80% during monsoon flow as runoff. The rapid increase in population, urbanization, and industrialization has led to a significant increase in water requirement in both purposes ultimately faced water crises. In the coming decade, the demand in water is expected to grow by 20%, fuelled primarily by the industrial requirements which are estimated to almost double from 23.2 trillion liters at present to 47 trillion liters of water. Domestic requirement of water is getting more as it was seen that nuclear society needs separate set up for family and water demand was also increased and its demand was increased by 40% from 56 trillion liters. Irrigation water demand in agriculture was increased by 14% which was estimated to be 592 trillion litters. Ministry of Water Resources of India showed per capita water availability in 2025 reduces up to half. As per the Ministry of Water Resources, industrial water use in India stands at 50 BCM. It is estimated about 6% of total freshwater use in industrial sector. This demand of fresh water is increasing and is expected to increase dramatically in the coming decade. About 90% of drinking water is supplied from groundwater and it is more hygiene compared with river water. As per the availability of water resource, it may used in different sector as per the utility of water. About 432 BCM of groundwater is utilized every year from river and rain water runoff. Out of this, 82% goes to irrigation and agricultural functions. On an average, the total groundwater present is about 10,812 BCM. Groundwater is decreasing at decreasing level day by day; use of groundwater is gradually increasing where groundwater recharge is decreasing. Groundwater recharge rate varies from region to region because of soil type; sandy gravelly soil promotes more water recharge compare with heavy soil. River basin recharge rate is about 250 m³ per day. Human being is more dependable on groundwater and surface water. Since the last two decades, report has shown that industrial water demand was most and it is increasing of the area of agricultural activity which needs fresh water. It is evident that the per capita availability of water is reducing progressively from 1816 m³. In view of increasing population per capita, water use will increase from 99 to 167 L per day in 2050. As per the database, the gap is increasing year after year and population growth is also increasing. The water demand is also increasing as per National Commission on Integrated water Resource Development Board in the domestic, agriculture, and industrial sectors. It was observed that water demand presumption made which may increase in all sectors and water demand may increase from 130 to 175 L per capita per day by 2025 and 2050, respectively.

Water demand may increase from 66 to 104 m³ per head by 2025 and 2050, respectively. A sum of water requirement may reach up 95–165 BCM by 2025 and 2050, respectively. World scenario on water purity and quantity revealed that in the near future, it may face tremendous water crises. Moreover, river water utilization may be restricted as it will show poor quality of water. Nation should go forward as increasing campaigns by Non-government Organizations and civil societies; awareness of water-related crises is increasing in every sectors.

The water requirement is increasing due to population explosion. It is true that the growth of population will never stop; it is a natural phenomenon. Once upon a time, very few human habitats were there around the world, now they are increasing with increasing population. Side by side all needs were fulfilled by modern people and human being faced global warming. The gap between the demand and supply of water is victimized by many countries in the world and consequence is given under:

a. Unavailability of safe drinking water
b. Unavailability of water for sanitation
c. Unavailability of water for waste disposal
d. Groundwater overexploitation leading to diminished agricultural yields

e. Misuse of water resources and adverse effect on biodiversity
f. Manmade pollution results poor quality water supply
g. Water distribution confliction within the country
h. Water-related to health issues
i. All the living beings will suffer due to water crises
j. Judicious use of water and its knowledge is very limited in human community.

India as well as the whole world is facing severe water crises in present days as India consumes more water as compared with other countries. By borne people are habituated to use plenty of water which results in wasting of water. Indians are gifted with many rivers in their terrain, naturally people generally use lots of water in every sector which is maximum compared with the other countries of the world. People were reported to have health hazards in many parts of the country due to lack of safe water. Judicious use of water is an urgent need in all sectors and we need more and more water-saving techniques in agriculture and other sectors. Besides, rain water harvesting strategies may help the nations. Without water, no function will complete, wasting of water is a criminal offence. These days, technology has been developed to save water in different sectors where efficient in every action is without water. However, to live everyday life, a minimum amount of water is needed but the variation is too high across the world; some regions face water scarcity and they should use water judiciously but they are utilizing more water in different ways instead of saving.

2.4.1 WATER AVAILABILITY IN INDIA

About 19 major river basins are present in India which are at different locations. The per capita water resource availability of these basins ranges in the Sabarmati basin is low about 240 m^3 and high about 17,000 m^3 in the Brahmaputra basin. In this area, water lift varies from 243 m^3 (Meghna basin) to 1670 m^3 in the Indus river basin. Irrigation water management for crop production is the main area of water. Water use in agriculture in this basin is to high compared with other basin. Western river basin is a low water production basin, that is, Luni river and its contribution is less than 6%. Indus and Pennar river basins stands second water resource basin and these river basins have significant food surpluses as these areas have high crop productivity. The grain surplus of the Indus basin alone is sufficient to meet 85% of the food grain supply from these basins. Third group of 11 rivers stands for about 75% of the population where crop production is in deficit compared with the other river basins. The water scarcity problems of the third group of 11 river basins home to 75% of the Indian population are mixed, but almost all have significant deficits in agricultural crop production. The fourth groups of river basins are classified as non-water scarce and food sufficient whereas the fifth groups of river basins are classified as non-water scarce and food surplus (Table 2.2). These last two groups of basins exposed more annual average rainfall and are regions of around 12% of India's population.

TABLE 2.2   Water Resources in India.

Water status	Km^3
Amount of surface water resources	1880
Amount of usable surface water	690
Amount of usable groundwater	418
Total usable water	1108
Water demand in 2002	780
Water demand in 2025 (estimated)	1200

As of now, India in the world having plenty water resources as evident from demand and supply of water resources in the country as earlier discussed. Water resources demand is lesser than supplies in almost all parts of the country except few states, that is, desert, red and lateritic soil, sandy soil and gravelly, and stony soil abundant. But water scarcity became evident from the frequent occurrence of drought at different locations of the country mainly due to lack of proper conservation and maintenance as well as uneven distribution of annual rainfall. As it was not possible to harvest excess rainwater during heavy to medium peak rainfall period, either that caused the disasters in the form of flood or flown to the sea and ocean or followed by drought in the period when there is no possibility of occurrence of rainfall. The aberrant weather leads to rainfall diversity zone to zone also creates a problem of proper water harvest. Hence, both the rainwater and irrigation water are considered to be equally important in the context of agricultural development of the country to get higher water use efficiency (Zaman et al., 2016).

2.5   STATUS OF WATER RESOURCES

The way of storing and conserving water is in underground, where about 98% of the entire world's water is in liquid form and also as fresh water and drinking water. Rest 2% presents as conserving water in river and lakes. Indians used more water per capita per head and not deprived in agricultural field also. The water loving crops are one of their options in cropping program. Best utilization of water is by the use of high water requiring crops. As per water resource availability, crop is selected for cropping schedule. It is uneconomic as water is a key factor for the present day agriculture. Water resource availability is in many forms, that is, surface water source, sub-surface, and reservoirs (lakes, pond, watershed, etc.) Every year, groundwater recharge occurred during rainy season and is utilized in dry season for agriculture purposes. The recharge rate depends on the pattern of rainfall, runoff, stream flow, permeability of the soils and bed rock materials. The total water resource availability per annum is 400 m ha m. Out of this, 180 m ha m is surface flow throughout the year. Out of 400 m ha m, about 105 m ha m comes from annual rainfall (Table 2.3). The per annum average groundwater recharge is about 67 m ha m, out of which about 50 m ha m comes from precipitation (rainfall). Hence, there is enough scope to increase the amount of rechargeable groundwater adopting ways and means. Mono cropping of rice uses huge quantity of water and three to four crops are taken per annum whether water is available for paddy cultivation or not. If not, shallow or deep tube well water is readily available. There are many means to take water from the any depth of under surface. The intensive use of groundwater results in water hazards and severe lower down of water table.

In India, natural process of groundwater recharge may be enhancing with many water conserving strategies. Water conserving strategies already taken in suitable manner but in addition to this, many things have to be included under water and soil conserving structures. Watershed development enhances the groundwater recharge. During rainy season, the aim of water and soil conserving structure should be more prominent. Minimizations of runoff have to be checked as much as possible. High rainfall areas of hilly terrain in rainy season is the maximum possibilities of groundwater recharge; gravel soil, porous sandy, and red and lateritic soil are good for groundwater recharge. Open wells and ditch are very common groundwater recharge in Odisha, Rajasthan, and Gujrat state of India during rainy season. Water infiltration rate observed maximum in high rainfall and porous soil. Heavy metal pollution occurs in high exploitation areas during dry season which is very dangerous for the whole humanity particularly arsenic, cadmium, and lead. Ground is more risk compared with groundwater in heavy metal contamination. Water table stabilization is a very important objective for our livelihood and emergency need of groundwater, and most of the areas are facing serious problem regarding lowering of groundwater table in that areas and may face severe water crises. Water resources for agricultural use, that is, irrigation are given in Table 2.3.

TABLE 2.3   Water Resources of India.

Annual average precipitation	1140 mm
Total available water	400 million ha
Net area sown	145 million ha
Gross cropped area	175 million ha
Irrigated area	70 million ha
Water demand for irrigation	46 million ha

2.6   IRRIGATION POTENTIAL

Irrigation potential of the country was reportedly created for about 22.6m ha. Food grain production of India is 50 million metric tons in the year 1951 and that reached 96.9 m ha having food grain production of 200 million tons resulting per capita food grain production is about 200 kg/year/person. The gap between irrigation potential whatsoever created at the huge investment of national exchequer and its utilization started from the year 1960s, is also becoming wider and that was serious concern of the researchers, policy makers, and high level planners for giving proper shape in future active integral part in domestic as well as industry and agriculture sector. Irrigation potential only can develop through different strategies which stand in front rows to meet up the human and animal needs. In many parts of India, life-saving irrigation water is very much essential for taking second crops, that is, rice-fallow areas and may expand the areas where *rabi* crops is possible.

2.7   IRRIGATED AGRICULTURE

The effect of supplemental water application as irrigation for crop production is a key to many places. Crop plant needs one or two life-saving irrigation water which is not available in many places of India. The plant has life, so it requires water for growth and development to give us proper quality and good yield otherwise socioeconomic status will deteriorate to some level. The water is required for metabolic activities including solubilization, absorption, and translocation on the way of biochemical processes in photosynthesis, respiration, and transpiration. The objective of irrigation management is different where availability of water is abundant that of limited or restricted water supply. The former is associated with yield increase per unit of land, irrespective of amount of water used for the purpose, whereas the latter is to increase the crop yield per unit of water use. Water requirement of crops and crop response to water use vary with different physiological growth stages. The certain growth stages are more sensitive to water stress situation. So particularly, where water supply is scarce or limited, it is necessary to take account of the application of water to the critical growth stages with respect to moisture requirement to get higher water use efficiency. If the crop suffers from water shortage during this period, either quality or yield of the crop reduces drastically. Higher yield of any crops may expect under all required essential inputs which may play an important role and only this can be achieved through providing optimum soil water and nutrients.

2.8 WATER PRODUCTIVITY

Water productivity is defined as crop yield per unit consumption of water. The crop yield can be produce per unit consumption of water. Evaporation, transpiration, gross inflow, net inflow, percolation, and seepage, the important components of water productivity can be included increasing yield per unit of land area. There are many water saver strategies for crop production which is agronomic practice, choice of crop, and efficient use of water. Water efficient management practices help growing crop and identifying water-saving opportunities. The concept holds good where water supply is for the purpose of irrigation.

2.9 STRATEGIES TO IMPROVE WATER PRODUCTIVITY

Groundwater Exploitation and Recharge: In this context, it was very much important to have the access as much as the excess water could be added to underground reservoir during the period of drought, would become easier and safer. There was clear indication that several soil conservation and water-harvesting measures could increase the groundwater recharge up to a certain limit. Groundwater recharge from open wells and pits are very common practice in Odisha, Gujarat, and Rajasthan. That is why greater exploitation of groundwater prior to rainy season provided scope of greater quantity to infiltrate during high rainfall period. Use of harvested rain water is safer than the underground source and only the flowing surface water is the safest source that is free from arsenic. So, all that an endeavor is undertaken by the planners and policy-makers and judicious use of water is the only way from ground without creating a problem which is directly or indirectly a health problem.

Agriculture accounted for at least 90% of the annual withdrawal of renewable water resource. Out of total 253 m ha irrigated land in the world, China and India alone have over 100 m ha. The available water resource declined rapidly due to lack of sufficient budget allocation for infrastructure development and degradation along with lack of proper maintenance of existing infrastructure for irrigation. Groundwater overexploitation is a major problem instead of judicious use of water, resulted in continuous dropping of water table with possible resultant health and environment hazards effect. The per capita withdrawal of water resource of about 612 m³, is too high as compared only 298 m³ in South Korea and 462 m³ in China, where India, South Korea, China are having renewable water resource availability was 2464, 1452, and 2427 m³ per year person, respectively. The per capita availability of water resource declined by 40–60% in most of the Asian countries over 1955–1990 periods. The water resource declining rate is an increasing rate in the coming future.

2.9.1   STATUS OF WATER PRODUCTIVITY

There is lack of judicious water use in agriculture from satisfactory output. Water productivity increment is important in water management for food security and sustainable agriculture. Water used by crop plants in an efficient manner results in higher crop yield. Excess water is very much detrimental to the crops because most of the soil nutrients may be lost through leaching process. Water requirement is determined by different types of crops; some of these are low water requiring crops, that is, pulses and oil seeds crops, and some of these are water loving crops, that is, rice. The water consumption has been increased for agriculture to 800 km³ and world population is 6000 million. The agricultural land is not increasing but the arable land is decreasing due to population explosion and invasion to arable land for the expansion of industry and human habitats. This can be summarized that an additional 800 km³ of water resources must be able to feed an additional 3000 million people in the world. The per capita water availability is estimated to be 0.72 m³ per day where world average is 2.4 m³ per day per capita for drinking and household purposes. The total water availability has to accommodate to this available water resource in the coming future.

It is estimated that per capita water needs for food production is 6 m³ per day. Water requirement is very crucial for the production food crops; whole world is feeding on that and survive. Most of the foods contains 80–90% of water and it is sometimes perishable in nature. Most of the fresh fruits and vegetables are perishable except stored grain food, that is, wheat, barley, rice, maize, pulses, etc., contains about 14% of water which is stored for long time. For an example, water requirement for the production of 1 kg rice is about 5000 L of water.

Water resource in the world is plenty; utilization of it in judicious manner would have sustainable agriculture and environment friendly. Human being is calling danger with misuse of water and polluting water; side by side polluting human organ causes several health hazards. Water use in agricultural sector and agricultural gains in future may be categorized in different components:

    i. Excess use of water in agriculture areas results high soil degradation, groundwater resource depletions.
    ii. Available water resource enhancement in rural segments.
    iii. Creation of enrichment of farming system.
    iv. Water conservation and limited water resource use.

2.9.2 *PRINCIPLES OF WATER PRODUCTIVITY IMPROVEMENT*

The following point focuses on improving water productivity in farm field, agricultural field, and basin areas:

    i. Crop yield enhancement per unit of water use.
    ii. Reduces losses of water, that is, seepage, infiltration, and runoff.
    iii. Enhancement of effective use of rainfall, conserve water, and water quality.
    iv. Proper use of water conveyance.
    v. Time of application of irrigation water.
    vi. Irrigation provides as per the need of crops and crop water requiring nature.

Water productivity enhancement may be practiced at farm field in which it improves in soil, water, and crop production management. There are many segments come under these are as follows:

    i. Crop selection
    ii. Variety selection
    iii. Planting geometry

   iv. Method of sowing and planting
   v. Types of tillage
   vi. Scheduling of irrigation
   vii. Nutrient management
   viii. Weed management
   ix. Plant protection

Many ways to soil water loss and depletion of water occur when water evaporates from soil and before crop establishment at early stage of crop. Agronomic management reduces soil water loss in such a way that different row spacing maintenance and the application of mulches improve water productivity strategies. The method of irrigation also checks the evaporative losses. Water losses can also be minimize by drip and sprinkler irrigation. Integration of crop and resource management also improved nutrient management strategies that can also increase water productivity. Integrated weed and pest management have also contributed effectively to yield increases as well as minimizes crop water loss.

Water productivity increment under deficit irrigation meets up the less water demand of crops. Under deficit irrigation crop may face severe water stress conditions and yield of crop may reduce. But lifesaving of water gives expected yield of crops where irrigated crops get more use of water instead of rain-fed crops. Under deficit irrigation farmers provides one or two irrigation for their crops and this crop selection is important to that zone because short duration pulse and oil seed crops selection is the criteria for production. Different irrigation methods depend upon the land situation and topography. In hilly areas, sprinkler irrigation is suitable whereas in orchard crops, drip irrigation is suitable for increased water use efficiency.

2.10   WATER FOR RICE CULTIVATION

The low water productivity per unit of water supply in rice production always checks the deep water percolation. Choking up the under surface of soil layer results in impervious layer. It obstructs the water to deep percolation or infiltration beyond the root zone. The main aim of this practice is to keep water retained for long time. Rice cultivation followed in clay to clay loam soil where water holding capacity is much more compared with the sandy or gravelly soil. Rice water productivity is compared in vivid manner with that of a dry cereal crops or coarse cereal grains crops, that is, jowar, bajra, *kodo, kutki, ragi, sawan*, etc., crops. Water saving technologies such as saturated soil and alternate wetting and drying results in increased water productivity. System rice intensification (SRI) is a technique of rice production which is reported for the first time in Madagascar. The SRI technique always proves increase in the productivity. SRI increases water productivity without sacrificing yield and quality grain production. Similarly, aerobic rice cultivation is possible without flooding of rice. This technique is followed in many places of India and abroad. The aerobic rice systems generally practiced in upland ecosystems and hilly areas. Irrigation water resources in rainfed agriculture are often related to uneven rainfall compared to low rainfall. The high risk for meteorological droughts are the results of uneven distribution of rainfall. Maintain yield effects lifesaving water that is required during dry spells and increases both production and water productivity dramatically if water is applied at the moisture sensitive stages of crop growth. Quite a substantial amount of water is added to irrigation water resources from rainfed ecosystem.

KEYWORDS

- **stakeholders of water uses**
- **agriculture**
- **domestic**
- **bovine**
- **power generation**
- **navigation**
- **industrial uses**

CHAPTER 3

Classification of Soil Water

ABSTRACT

Water resource development and management is intimately related with soil water being used in agriculture as agriculture and is an important and very pertinent stakeholder of water resource utilization. Water present in soil pores is called soil water. Soil water is an important component which influences soil organisms and plant growth and serves as a solvent and carrier of nutrients for plant growth and development that regulate soil temperature and helps in chemical and biological activities of soil. Those are also essential for soil forming processes and weathering. The retention and movement of water in soils, its uptake and translocation in plants, and its loss to the atmosphere are all energy-related phenomena. The more strongly water is held in the soil, the greater is the heat (energy) required. In other words, if water is to be removed from a moist soil, work has to be done against adsorptive forces. Conversely, when water absorbed by the soil, a negative amount of work is done. The movement is from a zone where the free energy of water is high (standing water table) to one where the free energy is low (a dry soil) is called energy concept of soil water. The difference between the energy states of soil water and pure free water is known as soil water potential.

3.1 INTRODUCTION

A water resource is entirely related with development of earth surface which corresponding to the three states of matter (solid, liquid, and gas) constitutes the earth. The solid zone is lithosphere; land which is covered by water forming seas and oceans is the hydrosphere. The gaseous envelope over the earth's surface is the atmosphere. The lithosphere consists of continents, oceans basins, plains, plateau and mountains, valleys, sand dunes and also it includes the interior of earth which consists of rocks and minerals. It is covered by gaseous and watery envelops. It accounts to 93.06% of the earth. The earth ball consists of three concentric rings: crust, mantle, and core. The crust is 5–56 km thick and consists of rocks with density of 2.6–3.0. It varies from 5 to 11 km in the oceans and 35–56 km in the continents. The soil scientists are interested in this skin.

3.2 SOIL WATER CLASSIFICATION

When water enters into the dry soil by rain or irrigation water, it is browsed around the soil particles, and it is held by adhesion and cohesive attraction. It replaces air in the pore spaces and pore fills with water. When all the pores, large and small are filled with water, soil is said to be saturated and it is at its maximum retentive capacity of the soil.

Soil moisture cannot be sharply stratified and differentiated from the clay particle and as per the uptake of water it is divided into following classes:

- Hygroscopic water (−31 to −10,000 bar)
- Capillary water (31 bar to 1/3 bar)
- Gravitation water (0 bar)
- Water vapor

3.3 HYGROSCOPIC WATER

Hygroscopic water is present on the surface of the soil and held very tightly on the surface of soil particles in very thin film by adsorption forces that act as adhesion and cohesion. It is mostly present in the soil as vapor form. The forces held on the surface are estimated to be about 10,132.5 bars toward the inner side and about 31.41 bars at the outer side. One atmosphere is about 15 pounds per square inch at sea level, which refers to the force present at one atmosphere is equal to about 14.6959 pounds per square inch or 1023 cm of water column height. This water is not available to the plants.

3.4 CAPILLARY WATER

It is the water control by the attraction of surface tension and continuous film around clay minerals and pore spaces. Once soil particles absorb water even when the hygroscopic coefficient is reached of the particular soil particle and thereafter more water is also held around the particles with fine layer. The water retains with fine layer and it is retained until the micro space is filled with water of the soil. When gravity force becomes stronger and additional water is also absorbed by soil particles, free water flows down.

The capillary-based water is present in the excess of hygroscopic means of water but is up to the point where the gravity attraction starts to move the water downwards, when free drainage conditions exist in the soil. Capillary water is loosely held water from 31 atmosphere to 1/3 atmosphere tension and is capable of movement within the soil particles. The plant nutrients are easily dissolved in it, and therefore, it is the most beneficial water for plants as the source of plant nutrients. The capillary water can be divided into two parts, these are given below.

3.4.1 *INNER CAPILLARY WATER*

It is some part of capillary water, which is close to the hygroscopic moisture (water) and is in the form of a thinner form of water, present more tightly and runs very slowly than outer capillary water as the water present in finer particle and the next stage of this is formed as vapor.

3.4.2 *OUTER CAPILLARY WATER*

Outer capillary water is not very tightly held in the clay minerals, and hence, thereafter runs readily from capillary tube to capillary tube. It is the most useful or available water to plants as it is very quickly absorbed by root hairs of the plants. A finer soil texture, granular structure, and coarse soil particle show larger pores. The single sand particle gives macropore as compared with clay soil particles having micro pore and large numbers results in more water passes in the sandy soil. Organically enriched soil presents organic biomass which gives macropores in the soil which results in high soil water holding capacity and capillary water.

3.5 GRAVITATIONAL WATER

Gravitational water is a type of water which is regulated by force of gravity and it moves freely in the soil. Once the maximum capillary capacity (MCC) of a soil gets saturated, variance in movement is by force of gravity. Free water moves between the soils particles through macro pores is called gravitational water

It is superfluous water and as such water plants cannot take. Gravitational water is present at zero atmosphere tension. Some adverse effects observed in gravitational water include soil aeration is affected badly and the most important is plant nutrient which is leached out beyond the root zone. The gravitational water adversely affect the rhizosphere.

3.5.1 WATER VAPOR

It is the water which is present in the form of gas, this gaseous form of water in the soil atmosphere is not directly used by plants. Soil water vapor depends upon the soil moisture content, atmospheric humidity, and temperature. The trend of increasing of vapor is directly involved with increasing moisture, humidity, and temperature.

3.6 WATER ABSORPTION AND MOVEMENT IN SOIL

The water movement through the soil is called water intake. Soil water adhesion depends upon the soil type and water content availability in the soil. It is the expression of different factors including infiltration and percolation. Absorption and movement of soil water condition variable under different soil states, that is, wet, dry and semi dry and its absorption depend upon the water attained percentage surrounding the soil particles.

3.7 INFILTRATION

It is the process of water entry into the soil generally through the soil surface and is vertically a downward movement. Infiltration rate determines the amount of runoff over the soil surface. Infiltration refers to the entry of water and its movement toward down into the soil surface. Infiltration is a surface phenomenon activity of a soil.

3.8   INFILTRATION RATE OF THE SOIL

Infiltration rate refers to how fast water enters from the ground surface to the deeper soil. Initially, the infiltration rate is more (dry soil) but afterward, it decreases because the soil becomes saturated. The rate of entry of water from surface to the soil, infiltration rate is grouped into four divisions.

1. **Very slow:** Soils with less than 0.25 cm/h, for example, very clay soils.
2. **Slow:** Infiltration rate of 0.25–1.25 cm/h, for example, high clay soils.
3. **Moderate:** Infiltration rate of 1.25–2.5 cm/h, for example, sandy loam or silty loam soils.
4. **Rapid:** Infiltration rate is more than 2.5 cm/h, for example, deep or sandy to silt loam soils.

3.9   FACTORS AFFECTING INFILTRATION RATE

Soil surface compactness: A compact soil surface allows less infiltration whereas more infiltration occurs from loose and friable soil surface.

Impact of rain drop: The force of rain drop that falls on the ground is said to be impact of rain drop on the surface. Generally, the size of rain drop varies from 0.5 to 4 mm in diameter. The speed of rain drop is 30 ft/s when reached at the surface, and the force is measured 14 times its own weight. Rain drop force causes sealing and closing of pores resulting in low infiltration rate.

Soil cover: Soil cover with vegetation is much more adventitious for better infiltration rate. Soil surface with vegetative cover has more infiltration rate than bare soil because the sealing of pores is not occurring.

Soil wetness: Infiltration is less in wet soil compare to the dry soil. When soil full of water rate of movement seems slow where in dry soil infiltration is more and the pore spaces are full of air.

Soil temperature: Warm soil contains less water or moisture compare to the cold soil absorbs more water than cold soils.

Soil texture: Infiltration rate is more in coarse-textured soils as compared with heavy soils (clay). In coarse-textured soil (sandy soil), the number of macropores are more. The cracking caused in clay soil by drying up with high solar radiation also increases infiltration in the initial stages until the soil again swells up.

Soil depth: Light soil of shallow soils allow less water to enter into soil than too deep or heavy clay soils. Infiltration rate is significantly more where coarse-textured soil, organic matter rich soil, and large number of microspace are available. This variable character of soil can be controlled by agronomic management practices like soil and water conservation management practices. Agronomic management practices loosen the surface soil and facilitate infiltration.

3.10   SOIL PERMEABILITY

Soil permeability of soil refers to the movement of air and water to the soil. It determines how fast air and water move through the soil profile. The entry of water into the top layer results in lower layer saturation and subsequent slow or rapid movement within the soil. It depends upon the bore size distribution in the soil. The basic things of permeability is more the number of macropores, the greater is the permeability. Slow movement of water in subsoil layers is due to their compactness and low organic matter content. Deep-rooted plant increased permeability even in such subsoil layers. Fine-textured and soil with sandy and silt proportion increase permeability.

3.10.1   FACTORS AFFECTING PERMEABILITY

- **Micropores numbers:** More the number of macropores result higher is the permeability.
- **Aggregates of soil:** Soil aggregates play an important role for water passing into the soil profile, where larger the size of capillary space greater is the permeability.
- **Soil depth:** Depth of the soil depends on the soil layer structure and class. As per class, permeability differs with the depth, as the subsoil layers are more compact and have less organic matter.
- **Soil texture coarseness:** Soil texture coarseness is also part of it for increasing and decreasing permeability; in coarse-textured soil, permeability is more where fine-textured soil permeability is less.
- **Salt concentration:** Salt concentration adversely affected the permeability. Sodium salt concentration is high in water and results in dispersion of soil and thus reduces permeability.

- **Soil moisture status:** Permeability decreases as the soil becomes drier and increases when soil becomes wet.
- **Soil organic matter:** More availability of soil organic matter in the soil results in more permeability of the soil.
- The permeability depends on the speed of water into the soil is considered slow, if it is less than 2.5 cm/h, moderate permeability calculated if it is about 5.0 cm/h in speed. Year after year tillage practices on the same piece of land reduces permeability. The growth of deep-rooted crops, particularly tap root system, that is, pulses or legumes, grasses, and tress increases permeability. The permeability of soil varies with its water present in the soil and usually decreases as the soil becomes drier and compact because air enters into soil and reduces the permeability.

3.11 PERCOLATION

The downward movement of water through saturated or nearly saturated soil due to the forces of gravity is known as percolation. Percolation happens when water is under pressure or when the tension is less than about 1/3 atmosphere. The conditions of water at a depth stable in such a layer where water goes deep into the soil until it meets the free water table. Percolation studies are important for two reasons:

1. The process of groundwater recharge occurs when it completes percolation. Even water is the only source of recharge of ground-water, which can be again be profitably used through springs and wells for irrigation, sometimes seemed as artesian well in many places.
2. This process also carries plant nutrients like calcium, magnesium deep into lower layers and along with water depositing them beyond the root zone. This process is enhanced in sandy or open-textured soils, there is a speedy loss of water through percolation.

3.11.1 FACTOR AFFECTING PERCOLATION

i. **Climate conditions:** Rainfall received more in such area having more than evaporation; there will be appreciable amount of percolation. In dry region, percolation is almost negligible or less.

ii. **Nature of soil:** Soil with sand part or gravel part exists, and such soil allows more percolation as these occupy large number of macropores. The macropores are used as the main channels of the gravitational flow. However, clayey soil permits less water to percolate beyond the rhizosphere.

3.12   CAPILLARY WATER MOVEMENT

There are many reasons to cease the gravitational forces where the water moves in the form of thin or capillary film from a wet region to dry region. The movement goes through the finer or micropores and it continues until the thickness of moisture film adjacent to the soil particles is equal to both the regions. Capillary tube and its movement may be in all directions, viz. downward, lateral, and upward. Water moves from thicker film around the soil particles flows to thinner film. The gradient of moisture between the thicknesses of the film, the quicker is the capillary movement up to certain point. The capillary tube moves in all directions and the movement of water film also becomes slow and may cease in some situation.

3.12.1   IMPACT OF FORCES ON WATER MOVEMENT AND RETENTION OF WATER IN SOIL

3.12.1.1   IMPACT OF FORCES ON MOVEMENT OF WATER

The forces are as follows:

1. **Gravity tension or gravitational force**
    The movement of water due to gravity refers to gravity tension. Soil is in saturated state and the direction of such flow is downward. The large pores, that is, macrocpores serve as the main channels for gravitational flow.

2. **Capillary tension or capillary force**
    When the soil water is held by the forces of surface in the capillary spaces and around the soil particles refers to as capillary force. Under unsaturated soil conditions due to force of surface tension, soil water downward movement. The flow due to gravitational force has ceased, the water moves in the form of capillary film from a saturated area to unsaturated area.

3. **Vapor tension**
    Under the dry conditions of the soil, the movement of water will
    have the gaseous form or vapor form and it may take place to a very
    little extent from soil layers. During daytime, soil gets heated toward
    the cooler soil layers.

4. **Osmotic pressure**
    The movement of water occurs due to difference in osmotic pres-
    sure. This osmotic pressure takes place in the soil system and the
    situation is exception in only saline soil which has excessive salts.

The forces, that is, the gravitational and capillary forces are important
because their significance in the movement of water in the soil is more. As
soil and its related force and rest two are less importance, these two are
vapor transfer and osmotic pressure and are less important because of their
negligible significance in case of normal soils situation.

3.13 SOIL WATER RETENTION

Water in the soil is retained by means of the following three forces:

i. **Adhesion force**
    The force occurs on solid surface to the water molecules. Because
    of the force of adhesion, the resulting water molecules are attached
    to the surface of soil particles. There is a thin film of water which is
    tightly presented above the surface of the soil particles. Due to finer
    soil particles, it expands in greater surface area. This characteristic
    of the water film is held or retained more tightly around the soil
    particles.

ii. **Cohesion force**
    This force occurs between similar molecules. It is defined as the
    attraction of soil water molecules for each other present in the soil.
    When more soil water gets saturated, the cohesive force shows its
    effect. The freshly added molecules get attracted toward present
    water molecules. Cohesion force results in thickening of water film
    above the surface of the soil particles.

iii. **Soil colloids**
    The soil colloids are also similar with clay or humus particles. The
    water is held in the soil due to soil colloids, that is, clay or humus

particles present in the soil. Presence of humus varies from soil to soil and region to region. The water held in the soil is called soak moisture or imbibition of soil particles.

The level of humus in the soil depends upon the organic manure source supply in that soil; therefore, this characteristic may vary from soil to soil. Heavy soil contains finer soil which is more or greater aggregation and organic rich soil results in more attraction of water in the soil particles.

3.14 SOIL MOISTURE CONSTANT

Soil moisture contents under certain standard conditions are referred to as soil moisture constants. The soil moisture content under field situation is always changing constantly with time and depth of soil. Therefore, it is not static or constant. Nevertheless, the idea of soil moisture constants greatly favors in taking decision of irrigation in time with efficient use of irrigation water.

3.15 IMPORTANT SOIL MOISTURE CONSTANT

Once exercise soil water and its availability, some specific terms called as soil moisture constants are generally used. Soil moisture constant having some important and commonly used soil water terms are given in table (Table 3.1).

TABLE 3.1 Soil Moisture Constant and Soil Moisture Tension in Atmosphere.

Soil conditions	Soil types	Soil moisture constant	Moisture tension in atmosphere
Wet soil	Gravitational water	Maximum water	0.001
Moist soil	Available water	Field capacity	0.33 (1/3)
	Water held in micro pores	Wilting point	15
Dry soil	Unavailable water tightly held	Hygroscopic coefficient	31
		Air dry	1000
		Oven dry	10,000

3.16   SOIL MOISTURE CONSTANT PARAMETERS

3.16.1   *HOT AIR OVEN DRY WEIGHT*

Hot air oven dry weight is the basis for all soil moisture calculations widely adopted at research station and university level. The sample soil is heated in an oven at 108°C until it loses no more water and then the final weight is recorded as oven dry weight in gram (g). The equivalent moisture tension at this phase is 10,000 atmospheres which means no water is there in the sample.

3.16.2   *OPEN AIR DRY WEIGHT*

Open air dry weight calculation is different from oven dry weight, but this is a variable constant. Soil exposed in humid conditions will have a higher weight. Normal soil sample without saturation contains less dry weight of the sample. Normal soil air dry at normal natural open air, the moisture is held with a force of about 1000 atmosphere. It also gives different moisture content during day and night. It also gives different moisture in different season, that is, humid, dry, rainy, or winter season.

3.16.3   *HYGROSCOPIC COEFFICIENT*

Hygroscopic coefficient is the capacity to hold maximum quantity of water by any soil sample in a saturated environment at 25°C temperature. Hygroscopic coefficient differs with the relative humidity percentage, temperature, type of soil, its texture, and organic matter percentage. The force of hygroscopic coefficient lies about 31 atmospheres and its constants are equal. It is determined by the soil in saturated conditions at 25°C temperature and soil moisture content at this constant is not available to plants. At this stage, the water presents in the vapor form between the soil particles but it is useful to certain soil bacteria and other soil microbes.

3.16.4   *PERMANENT WILTING POINT*

It is also known as a wilting coefficient or permanent wilting point or permanent wilting percentage (PWP). The moisture uptake pattern of roots of this situation is from the outer capillary regions, the roots hairs of the plant start to uptake with inconvenience in the inner capillary water. The moisture film becomes thinner in the capillary tube, the moisture present in the rhizosphere is held tightly and it is obstructed for the plant roots from browsing each successive portion of the water film present in it. Afterward, such a stage is reached in which plants cannot receive sufficient water for plant use and plants start wilting even after water availability, unless sufficient water is provided to the plant. The soil moisture constant at this stage is called as wilting coefficient or permanent wilting percentage. The force remains constant under 15 atmospheres at this stage or wilting coefficient and the wilting coefficient varied in different soils. In sandy soils, its value varies to 4–6%, whereas in clayey soil, it is high as about 16–20%.

3.16.5   *FIELD CAPACITY*

Field capacity   (F.C.) is the moisture content of a given soil sample present in % on oven dry basis, once it has been completely saturated and downward movement is ceased. Field capacity takes place after 2–3 days of high rainfall and artificial applied water and allowed to drain out the gravitational or free water. The moisture content at this stage in the soil is said to be at F.C. At this stage, all the pore spaces are filled with water. The F.C. is the upper limit of available soil moisture range in the soil which is helpful for plant. The moisture force at F.C. is about 1/3 atmosphere and it is constant. High organic matter content with fine-textured granular soil held more soil moisture than sandy soil at F.C.

3.16.6   *MOISTURE EQUIVALENT*

The moisture equivalents (M.E.) are the amount of moisture in percentage of a given sample on oven dry weight basis. The moisture percentage of given sample calculated in many ways, moisture on oven dry basis presented by 30 g of dry soil which is to apply 1000 times the gravitational force for half an hour in a centrifuge. F.C. may be assumed as equal to the M.E. and the value of moisture content may be considered as equal. The M.E. is a bit higher than filed capacity in sandy soil.

3.16.7   *MAXIMUM CAPILLARY CAPACITY*

When water is allowed to add to the soil whose FC is already reached and that water goes on thickening the moisture layer. A stage is then reached after which any further addition of water will get percolated down by the force of gravity. This is the point of maximum capillary capacity (MCC).

3.16.8 MAXIMUM WATER HOLDING CAPACITY

When the soil after its MCC is reached by the addition of water and it will start a downward movement by the force of gravity. While it is a well-drained out soil, but if drainage is restricted, the maximum amount of water can be presented until all pores are filled with water. This is called the maximum water holding capacity (MWHC). It is only in the case of poorly drained soils or soils having hardpan near the soil surface. Hardpan zone water remains stay for a long time. It is variable moisture constant because soil type is varied. The numerical values for this moisture constant for some types of soils are given below (Table 3.2).

TABLE 3.2 Moisture Constants of Soils (% Oven Dry Soil).

Soil type	Air dry moisture	Hygroscopic coefficient	Wilting Coefficient	Moisture equivalent	Maximum water holding capacity
Heavy black	3.8	20.7	29.9	53.2	79.7
Medium black	2.1	13.3	20.6	45.6	66.6
Alluvial	1.6	7.6	13.5	40.4	48.7
Sandy	0.5	1	5.3	21.8	25.2
Laterite	0.8	2.8	5.5	32.9	39.6

3.17 SOIL WATER AVAILABLE AND UNAVAILABLE

Available soil water indicates the availability of soil moisture by root hairs of the plants. The type of soil is crucial for keeping soil moistened for a long time for the plant to hold for a long period as long as the plant life cycle. Water present in the soil profile may not be available for the plants. The capillary tube and water present in the tube are assumed to be loosely held by the soil particles and water is not utilized by plants. Three divisions of the soil water are known on the basis of availability (Table 3.3).

  i.   Unavailable water
 ii.   Desirably available water
iii.   Superfluous or excess of water not used by plants.

TABLE 3.3 Available and Unavailable Water.

Type of water	Atmospheric pressure	Status
Oven dry	10,000	Unavailable water
Air dry	100	
Hygroscopic coefficient	31	Difficultly water
Wilting point	15	
Field capacity	0.33	Available water
Groundwater	0.001	Unavailable water

3.18 UNAVAILABLE SOIL WATER

The types of water which are not available to the plants are as follows:

a. Hygroscopic water
b. Fraction of inner capillary
c. Water vapor

a. **Hygroscopic Water**
    Hygroscopic coefficient and below of it is present so tenaciously above 31 atmospheres. At this stage, water is unavailable to plants. The water present between the hygroscopic coefficient and the wilting point is inner capillary water of soil particles. The circulation water is extremely slow or inactive and is only difficult to get available to plants. Certain types of plants under arid conditions make its use. The inner capillary water of soil particles can be used by some bacteria and fungi.

3.19 AVAILABLE OR DESIRABLY AVAILABLE WATER

Available water is present in between the limits of FC and wilting point (coefficient) is assumed as the available water. This water of the soil particles between FC and wilting point is the readily available moisture for all kinds of plants.

3.20 SUPERFLUOUS WATER

This water includes gravitational water and its excess of FC. Superfluous water cannot be absorbed by the plants. The soil water gets lost due to deep percolation beyond the rhizosphere. This superfluous water is harmful for plants if it stays for longer period. Under anaerobic conditions, most of the plants cannot take CO_2 and hamper the respiration process.

3.21 MOISTURE ABSORPTION BY CROPS

The absorption of water by the crops is related to the transpiration process. Water uptake depends on the rate of water loss in transpiration. The system may continue with at least once water is readily available to the rhizosphere. Water absorption pattern and transpiration rate are linked by the continuous water column in the xylem of plants. The expansion of water in the transpiration process promotes energy gradient which causes the uptake of water from the soil into the plants. Subsequently, it occurs from plants to atmosphere. With the presence of water column in xylem, the cohesive and adhesive properties of water play a significant role in the physiological process.

Water absorbs into plant roots by the process of osmosis, that is, transportation of liquid. This transportation occurs through semi-permeable membrane which is due to unequal concentration on both the sides. The concentration of soluble material in the cell sap of roots is accelerated because of loss of water through transpiration process. The concentration of soluble material in the cell sap of the plant is greater within roots than the soil moisture status. Moisture concentration present in rhizosphere equalizes the moisture concentration. The concentration of water molecule in the cell sap is reduced because of the quantity of soluble nutrients that occur. The water molecule is higher in number in the soil solution. First approach of water molecules in the root hair happens against cell sap.

Once the concentration of soluble nutrients in the soil water exceeds that cell sap, the situation will be stored and water will move toward the roots to the soil. In coastal ecosystems, plants grown in saline water with a high concentration of soluble salts absorb water with tremendous obstruction due to the high osmotic pressure of the soil nutrients. The water cycle is a part of the transpiration process, hence, the solar radiation provides energy for the vaporization of water from the water surface, soil surface, plant surface, vegetation, and plant leaves. The process of water loss from the leaf surface causes an increase in interior osmotic pressure. Due to interior osmotic pressure, water moves into them from the xylem vessels of the plants. The stem and roots of the plants cause a tension created by the loss of water from the leaf to be transmitted to the roots. The water uptake takes place in roots hairs and the maximum absorption takes place in the zone of root hairs.

Water uptake occurs mainly through roots hairs of the root of the plants. Water uptake occurs passively and actively with the physiological process. Passive absorption takes place when water is drawn into the roots by the negative pressure potential which is the phenomenon of transpiration. Under the situation during which there is little transpiration, the roots of many plants absorb water by expending energy that is called the active absorption process. In the usual state of transpiration, the contribution of active absorption to the water movement is minimal and it is usually less than 10% of the total uptake. Many plants communities are able to uptake moisture from the atmosphere when soil is at a permanent wilting coefficient which is un-utilizable to the plants. This phenomenon is also known as aerial uptake or negative transpiration. Simply, uptake of water by leaves that are saturated by precipitation (rainfall), dew, or overhead irrigation can help to re-saturate the dehydrated leaf tissue of the plants. The shoot portion including leaf and stem of the plant is mainly responsible for the loss of water. The leaf surface presents minute pores surrounded by two cells for the completion of transpiration. The plenty of pores present in the leaves are called stoma and stomata cells, and cells surrounding them are called guard cells. This cell, depending upon the environmental effects, facilitates the opening and closing of stomata. The stomata control the loss of water during day and night as vapor and exchange of CO_2 in leaves and other organs. The performance of this organ plays a vital role to determine the water loss from the plant. The performance of the stomata present in the leaves is depending upon temperature, humidity, air flow, size, and number per unit area.

3.22   SOIL MOISTURE ESTIMATION LABORATORY AND FIELD METHODS

Estimation of soil moisture within the root zones are done at regular intervals and at several depths. For estimation, information can be obtained as to the rate at which moisture is used up by the crops at different depths of the soil. For the estimation of soil moisture, key rules are established about when to irrigate and how much water to be applied and the time of application.

Irrigation may be given when about 50% of available moisture in the root zone is depleted. The quantity of water to be provided is directly related to the water already held in the soil. The methods of soil moisture measuring are given below:

A) **Direct method:** Finds out the moisture content in the soil.
B) **Indirect methods:** Finds out the water potential or stress or tension under which water is present in the soil.

A) Direct methods

I) Method of gravimetric

In this method, soil moisture measurement is made on soil samples of known weight or volume. Soil sample taken from the suitable depths are gathered in such a manner for even proportion with a soil collecting auger or soil sampler. Soil samples are taken away from bund, any depression, and organic source like organic matter present from desired depth at many locations of each soil type or same field for ideal average soil. These are stored in air tight aluminum containers to avoid iron containers to be free from rust. Later on, the soil samples are weighed and then dried in an oven at 105°C for about 24 h until all the moisture escapes off to complete the drying of soil sample. The dried sample is removed from the oven; after getting cooled slowly to room temperature and at which temperature the sample is weighed again. The difference in weight is the amount of moisture present in the soil. The moisture content (%) in the soil is estimated by the following equation:

$$\text{Moisture content (weight basis)} = \frac{\text{Wet weight} - \text{Dry weight}}{\text{Dry weight}} \times 100$$

PROBLEMS: Wet weight of a soil sample with can is 200 g and weight with can is 170 g and weight of empty container is 40 g. Calculated moisture content of soils sample?

Solution

Weight of wet soil sample = wet weight – weight of empty can
$$= 200 - 40$$
$$= 160$$

Dry weight of soil sample = Dry weight – weight of can
$$= 170 - 40$$
$$= 130$$

$$\text{Moisture content (\%)} = \frac{\text{Wet weight of soil} - \text{Dry weight of soil}}{\text{Dry weight of soil}} \times 100$$

$$= \frac{160 - 130}{140} \times 100$$

$$= \frac{30}{140} \times 100$$

$$= 21.4\%$$

II) Method of volumetric

The volumetric method is also convenient for moisture estimation; hence, the soil sample is collected with a core sampler or with a tube auger as per proper rules without hampering any step of sample collection whose volume is known. The main objectives of measuring the moisture from the soil should be in a proper way and the amount of water present in the soil sample is estimated by drying it in the oven and calculating by the following formula.

Moisture content = Moisture content (%) by weight × Bulk density (%) by volume.

Problem: For moisture estimation, the soil sample was collected from a field in such a situation where 2 days after irrigation when the soil moisture was near F.C.. The capacity of the sampler was 7.5 cm diameter and 15 cm deep for taking soil. The sampler weight of the core sampling cylinder was 1.56 kg and now estimate the available moisture-holding capacity of the soil and the water depth in centimeter per meter depth of soil?

Solution:

Moist soil weight	= 2.76 − 1.56 = 1.20 kg
Oven dry soil weight	= 2.61 − 2.56 = 1.05 kg

$$\text{Moisture content \%} = \frac{1.20 - 1.05}{1.05} \times 100$$

$$= 14.28\%$$

Core sampler volume $= d^2 \times h$

$= 7.5 \times 7.5 \times 15$

$= 662 \ cm^3$

$$\text{Apparent specific gravity} = \frac{\text{Weight of dry soil in grams}}{\text{Volume of soil in cubic centimeter}}$$

$$= \frac{1.05}{662} = 1.58$$

Available moisture = Approximate Specific Gravity × moisture content
= 1.58 × 14.28
= 22.56 cm/m depth of soil

The process of moisture calculation is accurate, convenient, and simple. It is generally used for the experimental purpose in research farm and academic institute. Many functions for estimation are sampling, transporting, and repeated weighing give errors. Moisture estimation is also a laborious and time-consuming job. The errors of the gravimetric method can be minimized with the help of increasing the size and number of samples. The sampling disturbs the experimental plots and hence many workers prefer indirect methods for the moisture estimation.

I) Using methyl alcohol

For the moisture calculation, soil sample is mixed with a known volume of methyl alcohol and then measure the change in specific gravity with a hydrometer. The way of calculation process may not be as usual but this method using methyl alcohol is a short cut procedure but it is not a common method in all the institute or organizations.

II) Using calcium chloride

By using calcium chloride for moisture estimation is enough scope but this method is one of them. The estimated soil sample is mixed with a known amount of calcium chloride for establishing the chemical reaction and water is the reactor and output gas measures the moisture of the soil. Then calcium chloride reacts with water under suitable water and temperature and removes it in the form of acetylene gas. The moisture is estimated has come in common use in the different institutes for research and academics.

B) Indirect methods

In these methods, no water content or moisture percentage in the soil is directly calculated but the water potential under which the water is held by the soil is measured. The most common instrument used for estimating soil moisture by the indirect method is given below:

1. Tensiometer
2. Gypsum block
3. Neutron probe
4. Pressure plate and pressure membrane apparatus

The data recording is done after the final results are obtained and the corresponding data of soil moisture content or percentage is measured by oven drying method. The moisture content as a percentage is depicted on a graph paper for easy access to the soil. Later on, these calibration graphs are used to knowns soil moisture content from the data of these instruments easily for understanding.

1. **Tensiometer**

   Tensiometer is also known as irrometer since they are used in irrigation scheduling for different crops as per the water consumption. Tensiometer supplies a direct calculation of tenacity with which water is present in soil. Tensiometer has a 7.5-cm porous ceramic or clay cup, a protective metallic tube, and a vacuum gauge installed in it. There is a hollow metallic tube holding all parts together. During the fitting of the instrument, the system is filled with water from the opening at the top and sealed with rubber cork and the set up placed in the soil below where moisture content will be calculated. Moisture from the cup moves out with drying of soil, creating a vacuum in the tube with certain pressure which is measured with the gauge. Proper installation is very much essential for actual reading in the active root zone of the crop. Gradually, when the desired tension is reached, the soil is irrigated and the tension is known for giving irrigation and as per the data, irrigation scheduling of the respective crop may be standardized. The vacuum gauge is graduated to indicate tension values up to 1 atmosphere and is divided into 50 divisions each of 0.2 atmosphere value. Under field condition and standardized set up, the tensiometer functioning is proper up to 0.85 bars atmosphere.

Advantages of tensiometer

1. It is very simple and easy to read soil moisture in field situations.
2. Scheduling irrigation to different crops which requires frequent irrigations at low tension is possible with the help of this instrument.

Limitations

Tensiometer functions well up to 0.85 atmospheres while the available soil moisture range is up to the atmosphere and hence is useful more on sandy soils wherein about 80% of the available water is held within 0.85 atmospheres.

2. **Gypsum blocks**
    A gypsum block is made by plaster of Paris. Its property and principles are used as resistance it is generally used for the estimation of soil moisture in situ. Gypsum blocks are commonly used in research institute. Gypsum blocks were first invented by Bouycos and Mick in 1940. The blocks are made of many materials like gypsum, nylon fiber, glass, plaster of Paris, or in a combination of these materials. The blocks are generally rectangular in shape. An importance of this unit is a pair of electronics which is generally made up of 20 mesh stainless steel wire screen solder to copper lead wire. Dimensions of screen electrodes are 33.75 cm long and 0.25 cm wide. The gap kept between the electrodes is 2 cm and a similar block is 5.5 cm long, 3.75 cm wide, and 2 cm girth.

3.23   WORKING PRINCIPLE

The important principle is electricity conductance. Once two electrodes A and B are placed parallel to each other in a medium and then electric current is passed through the electrode, the resistance to the flow of electricity is proportional to the moisture content in the given medium. Once block is wet, conductivity is high, and resistance is low shown in the unit. Generally, these read about 400–600 ohms resistance at FC and 50,000 ohms at wilting point. The readings are observed with a portable Wheatstone Bridge Bouycos water Bridge operated by dry cells.

During the placing of gypsum blocks in soil, care should be taken that the blocks must have close contact with the tightly placed soil. After setting the soil tightly with this unit, the blocks get saturated with soil moisture due to capillary movement. Pure gypsum block sets were kept for 30 min for the absorption of the moisture in the blocks. The gypsum block is very sensitive to the soil to moisture. The soil moisture tension is measured with limitations that varied from 1.0 atmosphere tension to 20.0 atmospheres. The gypsum blocks are suitable in too wet soils which is beyond the FC where Tensiometer is not working at all.

3. **Pressure membrane and pressure plate apparatus**
    Pressure plate apparatus for measuring moisture was developed by Richards. Pressure membrane and pressure plate apparatus

is generally used to measure FC, permanent wilting point, and moisture content or percentage at different pressures. The pressure plate apparatus consists of airtight metallic chamber in which a porous ceramic pressure plate is fixed tightly. The optimum pressure varying from 0.33 to 15 bars is applied. This apparatus consists of a compressor that supplies the pressure to the instrument. The measured water from the soil sample which is present at less than the pressure, applied trickles out of the outlet till an equilibrium against applied pressure is achieved after that the soil samples are taken out and oven-dried for estimating the moisture content in a given soil sample.

4. **Neutron meter**

   Soil moisture estimation is measured by the neutron meter method instantly. Neutron meter also gives continuous data without disturbing the soil. There is another advantage that soil moisture can be estimated from a large volume of soil sample. Neutron meter has a capacity to scans upto15 cm in diameter around the neutron probe in wet soil whereas in the case of dry soil, it is 50 cm. It consists of a probe and a scalar or a rate meter which gives the proper shape of that unit. The probe has fast neutron source, consists of radium and beryllium or Americium and beryllium. Another attachment with this unit is access tubes which are aluminum tubes of 50–100 cm length and are placed in the field where moisture can be measured. Neutron probe is the lowered part of this unit and inserted into the access tube to the desired depth of the soil. Fast neutrons are released from the probes, and then it scatters into the soil and gives the impact on data parameters. While the neutrons encounter nuclei of the hydrogen atom of water, their speed is gradually reduced. The scalar counts the number of slow neutrons, which are directly proportional to water molecules present in the sample soil, and its main objectives are to get the accurate moisture held in the soil. The moisture content of soil can be calculated from the calibration curve with counts of slow neutrons depicted in the neutron meter.

   Limitations

   There are two limitations of the unit. The first one is that it is expensive. Another limitation is the moisture content from shallow top layers cannot be measured with this meter. The fast neutrons are also slowed down by another source of hydrogen which is present in the organic matter. There are other atoms such as chlorine; boron, and iron also slow down the fast neutrons; therefore, it is overestimating the soil moisture content of the sample which is not desirable.

5. **Gama-ray absorption method**
   Gama-ray absorption method estimates the changes in soil water content by change in the amount of gamma radiation absorbed in different time interval. The amount of radiation passing into soil depends on soil destiny which varies mainly with change in water content or in percentage. Gama-ray absorption method is suitable where change in bulk destiny is very negligible.

6. **Feel and appearance method**
   For estimation of moisture content feel and appearance method is generally used. In this feel and appearance method, soil samples are taken from the desired depths. The soil sample is touched and squeezed in the hand and its feel and appearance are observed into consideration for the test. In this feel and appearance method, actual moisture content is not determined as given by the other measuring units.

7. **Soil moisture characteristic curve**
   An amount of water and energy of water in the soil is related with the soil moisture characteristic curve. Data base of soil water content is depicted into the form of a curve. The energy of water decreases with moisture toward more negative values where soil water content also decreases. As a result of the decreasing moisture content, more energy is utilized to absorb moisture from the soil. The relation between suction and water content of the soil is shown graphically with the help of a curve which is called as a soil moisture characteristic curve.

3.24   HYSTERESIS

The energy status and moisture content are classified into two ways:

   i.   Saturated soil sample and applied suction to dry the soil gradually and

ii. Slowly wetting an initially dry soil. The estimation of energy status
and moisture content are observed and plotted on graph chart. The
curves received through desorption and sorption are different.
Suction is greater in given sample in desorption than in absorption
and this characteristics is known as hysteresis.

3.25   TYPES OF WATER

3.25.1   *RAINWATER*

Rainwater is nothing but it is precipitated, harvested water in liquid form.
Various methods largely used to collect and store rainwater such as a roof,
land surface, or rock catchment. The water is mostly stored in a rain water
tank or directed to recharge groundwater by means of runoff obstruction.
Rainwater infiltration is another aspect of rainwater harvesting drawing an
important role in stormwater management and in the replenishment of the
groundwater levels. Rainwater harvesting was practiced over 4000 years
ago throughout the world, traditionally in arid and semi-arid areas, it has
provided drinking water, domestic water, and water for livestock and irrigation purposes. Nowaday, rainwater harvesting has attained much on significance as a modern, water saving, and simple technology around the world.

3.25.2   *STORMWATER*

Stormwater is the rainwater that flows across the surfaces, roofs, and paved
areas like roads, driveways, footpaths or yards, and flows to the stormwater
system. The stormwater system includes street gutters, drains, underground
pipes, and channels that transport rain water to water ways and groundwater.
Stormwater is not treated to remove some pollutants which irrigate the field.

3.25.3   *RIVER WATER OR LAKE WATER*

Rainwater running down according to the slopes ends up merging to form
small streams which then channel into a river. Due to the kinetic energy of the
moving water, the river develops various landforms through channel-making
processes. Rivers refer to any natural stream or waterways. Lakes refer to
any large body of water surrounded by a huge area. The basis of assuming

rivers and lakes is the hydrological cycle part of rain forming media. The hydrological cycle refers to the movement of water from the atmosphere to Earth's surface and ultimately to the oceans. Specific variations of the relative significance of the various components of the hydrological cycle estimate the volume of water held for rivers and lakes. The catchment collection guide to a river is known as the drainage basin. Drainage basin lithology and structural fabric determine the type of drainage pattern that a river displays. Sediment transport ability is strongly linked to relief. The total energy available for river sediment transport is the equivalent of a 7.5-kW engine removing material from 1 to 2 km for 24 h a day. Floods are shown to be an important part of the normal hydrological, geomorphological, and ecological functioning of a river. It is argued that rivers should be seen as an active system.

Lakes are connected with rivers through the hydrological cycle. Lakes have many origins; major lakes are usually formed by glacial erosion or through tectonic activity. Lakes are versatile and it has many use and lakes tend to be deep. Shallow lake with high levels of biodiversity during the middle of the 20th century that time dam building R accelerated to the point where over 65% of the world's rivers are now regulated. The impacts of this regulation are widespread and include detrimental hydrological, ecological, geomorphological, and socioeconomic effects. Year after year it develops a reverence for rivers and lakes, and recognize the interdependence of all biodiversity.

3.25.4   OCEAN WATER

The ocean water is dynamic in nature. Its weather parameters, that is, solar radiation, soil temperature, water salinity, water density, and the external forces like of the sun, moon, and winds influence the current of ocean water. The movement of an ocean in different directions, such as horizontal and vertical current is common in the ocean. The horizontal motion is known as ocean currents and waves occur season to season. The vertical motion refers to tides and their result of motion of the earth. Ocean currents are the continuous flow of a large amount of water in a specific direction while the waves are the horizontal motion of water. Water movement of the ocean is from one place to another through ocean currents while the water in the waves does not move. The vertical motion is to the rise and fall of water level in different heights at the oceans and seas. The result of the attraction of the sun and the moon according to the specific direction and the ocean water is raised up and falls down twice a day.

3.25.5   DOMESTIC WATER

Domestic water is supplied by the municipality, rural, and urban corporations for household uses. A good quality water source, treated water are required for households. The source of supplied water may be a stream, a spring in a bigger way like a piped water supply with tap stand or house connection, or water vendors. When it is a private initiation, it may be a hand-dug well, a borehole with a hand pump, and a rainwater collection system. Households use water for many purposes, that is, drinking water, cooking water, washing water, cleaning water, cooking utensils water, watering animals, and irrigating the garden. The supplied water may be different sources of water for different activities and the water sources available may change with the seasons. At someplaces, it is a very hard job to browse the water for dirking purposes and the source may be inadequate, however, it may be far away, difficult to reach, unsafe, or give little quantity of water, making it inaccessible or unavailable. These problems play an important role in people's health and well-being.

3.25.6   INDUSTRIAL WATER

Industrial water is effluent unsafe water and is one of the important sources in pollution of the water environment. A huge amount of polluted water was discharged into the river, lakes, and coastal areas resulting in serious water pollution problems and caused bad effects to all the ecosystem. Industrial water is of no use and it may recycle after treatment and its waste material may cause serious problems at different levels.

3.25.7   DRINKING WATER

Drinking water is defined as significant no risk to health over a lifetime of drinking, including much sensitivity that occurs between life stages. The greatest risk of waterborne diseases is for children, people living under unsanitary conditions. Safe drinking water is very much needed for all usual domestic purposes, including personal hygiene. The packaged water and ice intended for human consumption are maintained as per drinking water guidelines. Severely R may need to take additional awareness.

3.25.8   AGRICULTURAL WATER

Agricultural water is defined as the edible portion of a crop during growing, harvesting, processing, and packing. Utilization of water in different sector is the water used for irrigation of crops, pesticide, or fertilizer applications, preventing frost damage or crop dehydration or grain drying and washing or cooling of produce. Agricultural water is a potential way of contamination and by which pathogens can be spread to agricultural produce. Contamination of *E. coli* and Salmonella occurs in irrigation water in the production environment of fresh fruits, perishable products, and vegetables. Water quality should be pure otherwise it is dealing with fresh produce that is consumed raw. During the product arrival, it is difficult to separate pathogens from fresh produce and it is depending on the water contamination from the place of its transport. Many agricultural foods have rough surfaces that make them difficult to dislodge microorganisms from the surface during washing and cleaning. The enough washing is needed; even though enough pathogen is left on the contaminated produce which results in serious health hazards. Agricultural water is the water that is used to grow fresh produce, food grains, fruits, flower, and vegetables, spices, and livestock production. Agricultural water management includes the management of water used in crop production, livestock production, and inland fisheries in efficient manner. Stable and enriched production system is essential for sustainable water use and food security. Current food production may be doubled in order to meet the food needs of the world by 2050 by using the latest technology and judicious use of water.

3.25.9 IRRIGATION WATER

Irrigation is the artificial application of water to the plant. The effective irrigation has an influence on seedbed preparation, germination, root growth, nutrient uptake, plant growth, yield, and quality. The prime role to maximizing irrigation efforts is uniformity in the irrigation system. An important task for farmers is how much water to supply and when to apply irrigation. Decision taken up by grower in irrigation systems is best because your operation requires farm mechanization, varieties, growth stage, root structure, soil status, and land capability class. Irrigation systems should allow plant growth, once minimizing salt imbalances, leaf burns, soil erosion, and water loss may happen in the production process. Environment factors may affect the loss of water, such as evaporation, wind erosion, runoff and water nutrients, leaching, and seepage loss.

3.25.10 SEWAGE WATER

Sewage systems are underground pipes and the main lines which transport sewage from homes and wastewater from factories. Sewage water uses in different sectors like in agricultural fields for irrigation. System of treatment stage has to pass for getting quality water for irrigation purposes and other purposes which are very much needed to combat water crises in the near future. Sewage water is of no use; by approaching the modern water treatment process, this kind of huge water may be useful in the agriculture production system.

3.26   OTHER CLASSIFICATION OF WATER

3.26.1   WASTEWATER

Wastewater is the water which has been affected in quality by human use and agricultural use. Treated water may be used in agricultural field for irrigation, urban water use, and sewer inflow, and stormwater. Wastewater from a municipality is also called sewage and precipitation of this is called sludge. Water demands increasing day by day, this will become a much more common occurrence worldwide for many purposes. Industry must have appropriate approvals before discharging wastewater. It is important not to allow wastewater without treating it and before R it (Zimmer and Renault, 2003).

3.26.2   GREY WATER

Grey water also not in big scale but it has significant contribution for each and every house hold in many times. Wastewater of the output of clothes washing machines, showers, bath tubs, hand washing, lavatories, and sinks is called as grey water. This water generally is not in use except in few emergencies or when option is not available to fulfilling the need of that good quality of water.

3.26.3   VIRTUAL WATER

The fresh water availability per person having shrunk by two-thirds over the past 40 years, a new technology needs to be developed. Some eminents are of the opinion that a number of countries may be required to export water in the form of agricultural goods with virtual water embedded. Actual water is supplied from one country to another country because of lack of water of drinking purposes. The numbers will rest on those 31 countries which have come down from 62 that are projected to still have water in abundance in 2050.

The virtual water was reported in the year of 1990 and was first defined by Prof. Allan in the year of 1993 as the water is embedded in commodities.

Basically, water used to produce agricultural and industrial products is called the virtual water of the product. The third World Water Forum was held in Kyoto, Japan, and the gained popularity in the 1990s. The emphasis was given on water and added a new era to the debate on world water management. The highlighted phrase, "Virtual Water," can be traced to Israeli economists. Water is used for growing cereals, vegetables, meat, and dairy products. The amount of water used in the production process is called the "virtual water." The water in the contained product refers to virtual water. The "virtual" named, because it is not contained anymore in the product but it is transformed only. For the production of 1 kg of wheat, we need about 1000 L of water and for 1 kg meat production, we need about 5000 L of water. Nowaday rising global population and increasing water scarcity and human requirements for drinking and sanitation are pushing demand to scarcity levels. The daily requirement of drinking water is approximately 3 L, people consume "virtual water" in the form of food and fiber that require a huge quantity of water for their production. As per United Nations World Water Development (UNWWD) Report, people in developed nations have on average 3000 L virtual water consumption per day.

3.27 CONCLUSION

Soil water availability indicates the soil moisture by root hairs of the plants as an integral part of water resources. Soil water content is important for the plant to be held for long period as long as the plant life cycle. It is fact that all the water present in the soil profile is not available for the plants. Even the capillary water which is considered to be loosely held by the soil particles is not utilized by plants. Three tentative divisions of the soil water may be made on the basis of availability to the plant (agriculture) as: (a) unavailable water (b) desirably available water, and (c) superfluous or excess of water not used by plants.

TABLE 3.4 Virtual water available in agricultural products.

Sl. No	Products	Virtual water content (m³/tonnes)
1	Wheat	1160
2	Rice	1400
3	Soybean	2750
4	Beef	13,500
5	Pork	4600

Sl. No	Products	Virtual water content (m³/tonnes)
6	Poultry	4100
7	Eggs	2700
8	Milk	790

KEYWORDS

- **soil water**
- **hygroscopic**
- **capillary**
- **gravitational**
- **water vapor**
- **waste**
- **virtual water**

CHAPTER 4

Water Resource and Irrigation Command

ABSTRACT

India receives about three thousand billion cubic meters (BCM) of precipitation during the monsoon months, from the month of June to till October. The spatial variability is also very common as it varies between 100 mm in Western Rajasthan and 11000 mm at Cherrapunji in Meghalaya states. The rainfall pattern of India is so varied and the aberrant weather conditions lead to the variability of the rainfall. There are 13 major river basins in the country having a catchment area coverage exceeding 250 km^2. The country is embedded with some limitations of physiographic conditions, socio-political environment, legal and constitutional constraints, and the technology availability; the utilizable water resources of the country have been observed at 1123 billion cubic meters, of which 690 billion cubic meters is from surface water and 433 billion cubic meters from groundwater sources. Considering the availing of about 60% (i.e., of 690 billion cubic meters) of utilizable surface water is possible only if matching with existing water harvesting structures. Water supply from basin areas, considering to the full extent as proposed under the National Perspective Plan, would further increase the utilizable quantity by approximately 230 billion cubic meters. The irrigation potential of the country has been estimated to be 140 million hectare without interbasin sharing of water and 175 million hectare with interbasin sharing. There are four main sources of water: (i) above groundwater/surface water, (ii) undergroundwater/groundwater, (iii) atmospheric gas form water, and (iv) ocean water.

4.1 INTRODUCTION

There are many important issues raised under irrigation in agriculture that has been fixed on the basis of certain standard duties in terms of crop area irrigated per unit of water. The proposed and established cropping pattern is decided on the basis of past experience in the particular region. The design of the various component parts of the tank and farm ponds usually requires specialized engineering knowledge for accelerating water harvesting activities. The tanks or farm pond/reservoirs fail mainly due to two reasons: (i) silting of bed and (ii) breach due to inadequate over surplus arrangement like soil erosion or bad maintenance of the boundaries without vegetation that enhances the soil erosion. Renovation of long-term used tanks so as to restore the lost irrigation potential is being accorded priority under the minor irrigation program. The function of restoration generally consists of (i) strengthening or rising of bund, (ii) improving the surplus capacity, and (iii) occasional de-silting of bed. De-silting is expensive, but in some cases this is being rendered economical by utilizing the excavated earth for reclaiming part of the previously submerged land. After renovation of it, irrigation works below a specified acreage are handed over to the panchayats or local bodies for maintenance. Works having higher irrigation capacity are maintained by the Public Works Department as per the situation and status of the system.

4.2 CULTIVABLE COMMAND AREA

Cultivable command areas (CCA) are those areas which are feasible for irrigation facility as well as the well-cultivated land. The area can be irrigated from a scheme and is fit for cultivation as its facilities are available for livelihood security and sustainable agriculture.

4.3 GROSS IRRIGATED AREA

Irrigation opportunities are available for supplying water for a given time period under the farming community. The yearly covering area per crop under irrigation and counting the area irrigated under more than one crop during the same year as many times as the number of crops grown per annum and irrigated to the crop for optimization of yield.

4.4 IRRIGATION POTENTIAL CREATED

This is the scheme which is developed to provide facilities for supplying irrigation water under different crop enterprises. The total cropped area under the given area to be irrigated in different crops during a year. The area may have to be irrigated under more than one crop during the same year is counted as many times as the number of crops grown and irrigated under this system or scheme.

4.5 IRRIGATION POTENTIAL UTILIZED

This potentiality depends on the availability of the water for irrigations which can supply to the crops for life-saving irrigation. The gross area actually irrigated during the reference year out of the gross proposed area to be irrigated by the scheme during the crop year.

4.6 MINOR IRRIGATION SCHEME

MI scheme having CCA up to 2000 ha individually is divided as MI scheme in given area.

4.7 MEDIUM IRRIGATION SCHEME

Medium irrigation scheme refers to a scheme in which the CCA is more than 2000 ha and up to 10,000 ha.

4.8 MAJOR IRRIGATION

Major irrigation scheme is a scheme in which the CCA is more than 10,000 ha.

4.9 TYPES OF MINOR IRRIGATION SCHEMES

4.9.1 DUG WELL

Under this scheme, water is lifted from dug well to give life-saving irrigation. It covers ordinary open wells of varying dimension dug or sunk from the ground surface into water-bearing stratum to extract water for irrigation purposes on a small scale. These are generally masonry wells, earthen wells (age-old practice), and dug cum bore wells with dual functions. All such schemes are of private nature belonging to individual cultivator and the area coverage is less as compared with the other scheme.

4.9.2   SHALLOW TUBE WELL

Shallow tube well is quite popular and it consists of a borehole built into the ground with the purpose of tapping groundwater from porous zones. Due to the sedimentary formations, the depth of a shallow tube well does not exceed about 60–70 m. These tube wells are either cavity tube wells or strainer tube wells. These are usually mechanically drilled by percussion. Method using hand boring sets and sometimes percussion rigs depends upon the expert developed the tube wells locally. The success and popularity of the scheme depend on how cheap they are and their availability in the locality. Indigenous structures like coconut rope formed by binding coir strings over an iron frame are being used as a strainer and somewhere it is iron net or structure. In shallow water table areas, bamboo frames are also used for easy and cheap cost. Sometimes, steel casing pipes are replaced by pipes constructed by rapping bituminized gunny bags over the bamboo frame for durability. These are called bore wells in which the borehole is stable without a lining in the bottom portion and a tube is inserted only in the upper zone because it browses the water layer. The tube wells are generally operated for 6–8 h during the irrigation season and give a yield of 100–300 m³/day, which is roughly two to three times that of a dug well, and ultimately the efficiency is quite more.

4.9.3   DEEP TUBE WELLS

It is costly because it contains durable long-lasting deep iron or PVC pipe and it covers large area for providing irrigation. It usually extends to the depth of 100 m and more and is designed to give a discharge of 100–200 m³/h. The deep tube well is drilled by rotary percussion or rotary cum percussion rigs with the help of a pipe installation machine set. The tube wells work round the clock during the cropping season, depending upon the availability of power. The most important aspect is that the annual output is roughly 15 times more than that of an average shallow tube well and is usually constructed as a public scheme which are owned and operated by government departments or corporations for providing this facility.

4.9.4 SURFACE FLOW

Surface flow is generally adopted by farmers, government, and non-government schemes. This program consists of age-old technique or indigenous culture for providing irrigation to the crop fields. Use of surface flow water or rainwater for irrigation purposes is either by storing it or by diverting it from a stream, *nala*, or river, this system is developed as per human need in different times. In many times, permanent diversions are constructed for using the flowing water of a river. The structure like temporary diversions as per the feasibility is also created in many areas which are usually washed away during the rainy season. The small storage tanks are called ponds or *bundhis* which are mostly community-owned. The command areas under this scheme are about 20 ha or less than that. The large storage tanks whose command varies from 20 to 2000 ha are generally constructed by government departments or local bodies. These are the larger area expansion task of surface MI.

4.10 STORAGE SCHEMES

Storage structures may be including tanks, water-conserving structure. Water collects in small containers or cement concrete structures, ponds, farm ponds, and reservoirs which impound water of streams and rivers for irrigation purposes. Farm ponds, wells, tanks occupy a very important place under the MI activity. They provide irrigation of nearly two-third of the total irrigation only from minor sources in the states of Andhra Pradesh, Karnataka, Kerala, Maharashtra, Odisha, West Bengal, Jharkhand, Bihar, and Tamil Nadu. These states have plenty of water of river in which some are the Himalayan origin and some are the monsoon rain origin. Many of these places are not even land and mostly it is an uneven topographical distribution. Besides, there are scope for further construction of farm ponds and tanks in many areas. Many of these tanks in southern states have gone into not in use due to long neglect of repairs or renovation as siltation is a common problem in the entire storage scheme.

Repair or renovation of these farm ponds, tanks, or storing structure so as to restore the lost irrigation potential is being accorded advance plan. The crucial features of these schemes are as follows:

i.   Bund is made of earth, but is also sometimes masonry.
ii. Feeder channels to divert water from adjacent catchments area.
iii. Waste weir to drain excess flood water.
iv. Sluices to supply for irrigation.
v.   Conveyance and distribution of irrigation system.

The size of the storage for water is determined by the runoff expected on the basis of rainfall pattern in the catchment and by the fact whether the rainfall and cropping program. The best and direct method to calculate the surface flows may be to gauge the streamflow at the given site for a number of years. The runoff is computed on the basis of a formula where the past experience was established of the catchment area. Farm ponds are made in a series by bund up the same valley at several points; some spillover yield from the bund catchment is also accounted for. The total storage provided some percentage is permitted for dead storage to be utilized by silting with time.

4.11 DIVERSION SCHEMES

The main objective of this system is to divert the huge water by gravity irrigation of stream water supply without storage structure for the future use. There are many storage program and they are economical but their accessibility is performed on the presence of flow in the stream at the time of actual irrigation requirements of a particular phase of the crops. There are essentially such schemes consist of following points:

i.   Bund constructed at the bank of the stream for diverting the water for irrigation. The weir being called *anicut* in the Southern India, *bandhara* in Maharashtra and Gujarat states, and *thingal* in the Assam.
ii. Artificial water channel for irrigation which is popularly known as *kul* in the hilly areas, *pyne* in Chhota Nagpur and Bihar and *dong* and *ilhowkong* in the states of Assam. It is very popular small schemes which have ample scope in the hilly areas and foot hill plains of India.
iii. The water is usually diverted by constructing temporary bunds across the waterways, bund up of earth (soil or mud), stones, or even bamboos according to the availability of the materials at the locality. This indigenous discharge of water occurs in a small bund on the head of the channel which is not provided with any gated structure for controlling and regulating the flow.

iv. All the components are very innovative to accomplish the system which is simple and cheap and can be handled to a large extent at the farmer's level. The components being temporary require frequent renovation due to easily wear and tear.

v. The bunds are easily fragile to be washed away by every major flood and water pressure. The water channels also get silted up blocked and scoured frequently for getting better facilities. Important task of this structure is to divert a high volume of water, that is, more than 5–10 cusecs, or tackle a river having a high intensity of flood discharge. The other important aspects are structures equipped with suitable types of gates are given. Different types of weir have sluices which regulate the flow of silt in the off taking stream. The cement-structured weir is simpler and cheaper where hard stony bedrock is laid upon the stream. The structure of the weir depends on infiltration capacity and is constructed by a special engineer who has knowledge of it. The capacity of the water supply is dependent on the volume of flow in the river at the time the irrigation is required. Many important issues are concerned of snow bed streams in the rainy season, after which they discharge enough amount of water. A few diversion schemes are also constructed as monsoon channels supplying water only during the monsoon time. These schemes are useful for providing supplemental irrigation for paddy and preliminary watering for sowing in the winter season.

vi. Small irrigation is required in mountain agriculture systems and irrigation channels called "kuls" in hilly area are the only way of irrigation and are affordable cheap application of water to the field. These waterways carry water diverted from streams by constructing temporary or concrete bunds across the streams or waterways. It is very difficult to supply the water to the field from one place to another in hilly areas and most of the time unfavorable conditions occur during construction of channels.

There are advantages of water conservation through groundwater recharging steps: Water conservation and its many objectives are stated below:

i. Inundation of agricultural land during monsoon for sowing post monsoon crops,

ii. Development of moisture level of the adjoining fields downstream for raising of the second crops with develop moisture. Some advantages of these programs are that they help to conserve the soil.

iii. Water conservation in the catchment area of tanks down below serves the important purpose for retarding the silting rate. Water conservation through the field is peculiar in central Indian tracts is known in the northern Madhya Pradesh and Chhattisgarh. Water conservation in situ also developed in Bundelkhand area of Uttar Pradesh and Rajasthan. During monsoon, water is conserved as stored in upstream and the land gets submerged in a huge area afterward when water is required. Generally, no direct irrigation is carried out and the advantages are mostly due to the submergence of water.

iv. Moisture retention capacity of soil is categorized in some states of India. Such soil is black which is retentive of soil moisture results in enough amount of soil moisture conservation. During the monsoon, water retains in the soil and absorbs sufficient moisture to grow *rabi* crops. The excess water is used for the cultivation and also for different purposes. One of the advantages of the first flood is rich soil health that any crop can grow without fertilization. Free flow of water across steep of the soil to gain enough soil moisture, the soil of the land is also conserved for multi-uses. It is very common method or technique for preventing the water loss from the soil when water is stored in soil or conserving techniques. Water is let out in October month the drained out in the bed of the *Ahars which* are cultivated with winter crops and the term *Ahars* is most common in Bihar.

v. Bund is put up across the land slope at the head of gullies with the aim of impounding and diverting the runoff into the wider zone. Land surface irrigation channels are given in the flanks to carry flood water another season when sufficient rain is received. Groundwater recharge occurs in an efficient manner in Andra Pradesh, Maharashtra, Tamil Nadu, Kerala, Karnataka, and Rajasthan.

4.12   IRRIGATION POTENTIAL CREATION AND UTILIZATION

The total precipitation data shows that India receives about 4000 billion cubic meters (BCM) of annual precipitation which is a sufficient rainfall pattern reflected a very high degree of diversity in the different states. About 3000 BCM of precipitation occurs during the monsoon months from the month of June to till October. The spatial variability is also very common in Western Rajasthan and its variation is 100 mm whereas in Cherrapunji in Meghalaya state it is 1100 mm. The rainfall pattern of India is so varied and the aberrant weather conditions lead to the variability of the rainfall. There are 13 major river basins having a catchment area cover about 250 km^2 and more. Some Indian situations have the limitations of topographical conditions, social and political arena, legal and constitutional obstructions. The utilizable water resources have been noted at 1123 BCM. Out of 1123 BCM of which 690 BCM is from surface water contribution whereas 433 BCM is available from undergroundwater sources availability. Availing about 60%, that is, of 690 BCM, of utilizable surface water is possible. Water supply from basin areas may increase the utilizable quantity by approximately 230 BCM for the agriculture use. The supply of artificial application of water of the country has been figured out about 140 million hectare (m ha) without inter-basin sharing of water and 175 m ha with inter-basin sharing. There are four main sources of water:

    i.   Above groundwater
   ii.   Undergroundwater
  iii.   Atmospheric gas form water, and
  iv.   Ocean water

Water for the irrigation purposes water came to about 85% of the total water used. These days water demand is increasing for irrigation is may be fall to about 73% by 2025 AD. and different water use efficiency strategies has been taken in large scale. The irrigation potentiality has been again assessed at 139.89 m ha. The criteria based on the reassessment of the undergroundwater potential increased to 64.05 m ha. It was also looked back to the water potential which was 40 m ha earlier. Reassessment was done of the potentiality of surface for MI sector which was from 15 to 17 m ha. Thus, there has been an increase of 26.39 m ha in the ultimate irrigation potential, which was 113.5 m ha before re-assessment. Irrigation potential creation and its judicious utilization are the objectives of Government planning and program for effective utilization. Some involvement of irrigation, that is, micro irrigation for water management practices these are sprinkler and drip irrigation systems in water scarce and drought prone areas. Many knowledge sharing programme on conjunctive use of surface and groundwater and farmers' participation in irrigation water management programme for judicious use of water. With utilization of irrigation potential mainly under major and medium irrigation plan, continues to future use and improvement in the irrigation projects. To minimize the gap between the irrigation potential created and utilized and to uplift water productivity the Command Area Development Program (CADP) was implemented (Table 4.1). National Water Policy (NWP) during 1999–2000 states that the water rate should be such that farmers know the short fall of water in agriculture. Some irrigation-related initiations help in yearly maintenance and operation cost.

TABLE 4.1   Irrigation Potential Creation and Utilization.

Irrigation schemes	At the end of seventh plan	At the end of seventh plan	Ninth plan target	At the end of ninth plan target
Major and medium irrigation (mha)				
Potential	29.9	32.96	9.81	35.35
Utilization	25.5	28.44	8.71	30.47
Minor irrigation (mha)				
Potential	46.6	56.60	7.24	59.38
Utilization	43.1	52.32	4.93	54.23
Total (mha)				
Potential	76.5	89.56	17.05	94.73
Utilization	68.6	80.76	13.64	84.70

4.13   CAUSES OF GAP IN BETWEEN CREATION AND UTILIZATION

The irrigation demand increases as per crop per year increment. Irrigation potential created for agricultural purposes is the total area which can be irrigated from a plan on its full utilization. This important before an area is to be estimated under "potential created" for effective utilization.

1. The estimates of ultimate irrigation potential are relevant for a particular phase since the estimate is derived on the basis of a number of assumptions about cropping pattern and water allowance, which undoubtedly vary over time.
2. The estimate of total irrigated area is a possible under estimate as areas under the two seasonal and perennial crops are counted only once. However, it may be over estimated if areas under the other projects from the new command area are included.
3. Estimates of CCA are often arbitrarily arrived at without carrying out any survey.
4. IPC of a new project is the aggregate of all areas at the end of water-courses where water could be delivered from the project and IPU is the total gross area actually irrigated during the year which has been the base of estimation.

5. IPC and IPU are the parameters developed by the Planning Commission for monitoring a project and are to be compared in a project-specific manner. They, perhaps, cannot be aggregated at a regional level and compared.
6. There are possibilities in variations in estimates of IPC and IPU as different organizations compute them with different objectives.
7. The gap should be tried to be bridged through microlevel and farm-level irrigation water management techniques.
8. The gap is mainly attributed to magnitude of wastage, a mismanagement of irrigation water while the stakeholders are utilizing the precious resource like water being created at the cost of huge national exchequer.

4.14 REMEDIAL MEASURES TO MEET UP THE GAP

The large demand exists between gross-cropped area and gross-irrigated area not increased as of now, needs to be implemented for increasing productivity, production level, and resilience.

1. Development of microlevel water resources (tank-well system, small springs, perennial, and seasonal streams) with a command of 2–10 ha to reduce the dependency on major and medium irrigation projects through R&D projects should be exploited.
2. Effective harvesting of excess rainwater during heavy to medium rainfall period as well as artificial recharge of groundwater should fully be exploited wherever possible through participatory R&D and mass awareness program.
3. Planned intervention is required to reduce the negative effects of surplus exploitation of groundwater by two-fold actions: (1) to control excessive draw down and (2) to prevent water quality deterioration and degradation.
4. The gap between technology generation and adoption in the farmers' field are to be minimized; the project is contemplating with the idea of having Water Technology Park as live demonstration is to be planned. The methods, systems, and scheduling of irrigation based on different approaches, its transformation in farmers' language, water harvesting structures are to be undertaken.
5. The future of major irrigation command and medium irrigation command should be studied in depth and participatory water management research as well as on-farm research should be carried out.

6. Generation of suitable technologies on rainwater harvesting and recycling of excess water for its effective use for crop production.
7. The evaluation of suitable rice and other crop cultivars along with improved agrotechniques for the coastal saline zone.
8. The study on reclamation of salinity, use of saline water for irrigation, breeding of salt tolerant varieties of crops, and better agronomic practices suitable for Sundarban area for higher yield and effective utilization of saline soil.
9. Effective utilization of low land ecosystem or wet land ecosystem through integrated farming with rice-vegetable and fish components.
10. To mitigate the arsenic problem in water with suitable technology available to the farmers.
11. Approaches such as scheduling irrigation to crop and crop sequence with judicious and conjunctive use of rain/canal and groundwater keeping parity with optimum level of groundwater.
12. The study on groundwater fluctuations, its adverse effect on soil health as well as to evolve the methodology to raise groundwater table.
13. The study on the crops like date palm, swamp taro, colocasia, lettuce, water chestnut, and other in view to meet the ever increasing food demand.
14. Management of the horticultural crops in orchards, flowers, and vegetable to find out suitable alternative of rice crop and to introduce less budget high value crops.
15. Performance of the traditional rice varieties for its selection for suitable by-products as well as coproducts including indigenous technologies for the alternate use of rice.
16. To develop suitable agrotechnologies for the rainfed farming based on farmers' participation and introduction of crops with less water requirement.
17. To replace the summer paddy by other substitute crop(s) without hampering the economic benefits of the farmers.
18. To utilize the plant protection measures more efficiently by introducing IPM.
19. To introduce paira/mixed crop in risk prone areas to ascertain the farmers benefits more effectively.
20. To increase the area under the adaptive and innovative trials on several aspects of agricultural production enhancement activities based on participatory approach.
21. Development of submerged and mercy land (*jheels and beels*) for increasing agricultural productivity.

The country is having irrigation projects which are of multipurpose and major irrigation program. These projects were initiated for increasing water productivity, such as Bhakra Nangal, Nagarjuna Sagar, Kosi, Chambal, Hirakud, Kakrapar, and Tungabhadra river projects. Concurrently, ground-water was given priority under the agricultural sector along with the financial assistance from the Centre. CADP was launched with the objective of reducing the lag between the potential created and the optimum utilization of available land and water resources. All related activities comes under one umbrella because initially, 60 Major and Medium projects were frame out with a CCA of 15 mha. The greater emphasis was laid on the completion of projects, which were in advanced stages of completion. During the 8th plan period, irrigation potential of 2.22 mha was created under the major and medium sector at an annual rate of 0.44 mha per annum presented in Table 4.2. Some of the old irrigation schemes were renovated, modernized, and rehabilitated with high priority in the mode of repairs and improvement. The MI projects were part of integrated micro-development and they gained enough scope to the irrigation sector. With this view, sprinkler and drip irrigation programs and the conjunctive use of surface and groundwater gained momentum. There are ample scope of the plan-wise initiatives taken in augmenting the irrigation capacity in India, the study now proceeds to strengthen on the utilization of the irrigation potential.

TABLE 4.2 Plan-wise Irrigation Potential Created and Utilized (in m ha).

Plan period	Potential created	Potential utilized
Up to 1951 pre-plan	22.60	22.60
I plan (1951–56)	26.26	25.04
II plan (1956–61)	29.08	27.80
III plan (1961–66)	33.37	32.17
Annual plan (1966–69)	37.10	35.75
IV plan (1969–74)	44.20	41.89
V plan (1974–78)	52.02	48.46
Annual plan (1978–80)	56.61	52.64
VI plan (1980–85)	65.22	58.82
VII plan (1985–90)	76.53	68.59
Annual plan (1990–92)	81.09	72.85
VIII plan (1992–97)	86.26	76.27
IX plan (1997–2002)	93.95	80.06

It is estimated that the total cultivable area is about 186 m ha and net sown area is about 143 m ha. The country has ample scope to utilize this ultimate potential by the year 2018. The initiation has taken over the utilization of water resources of the country. The total irrigation potential has increased from 82 mha in 1991–1992 to 108.2 m ha in March, 2010 (Table 4.3).

TABLE 4.3   Development Irrigation Potential (m ha).

Year	Major plan	Minor plan	Total
1950–51	10	13	23
1980–81	27	31	58
1999–2000	35	60	95
2006–07	42	60	102
Ultimate potential	73	66	139

4.15   IRRIGATION POTENTIAL UTILIZATION SCENARIO OF INDIA

The irrigation potential development and utilization made by different major, medium, and MI projects. During eighties, the irrigation potential was about 58 m ha where the net utilization was only 54 m ha (Table 4.4);

TABLE 4.4   Shows the development and utilization of irrigation potential in India in recent years (m ha).

Irrigation plan	Seventh plan	Eighth plan	Ninth plan	Tenth plan
Major and medium irrigation potential	30	33	37 31	42
Utilization	26	29		35
Minor irrigation potential	47	53	57	60
Utilization	43	50	50	53
Total potential	76	86	94	103
Utilization	69	77	81	87

The total irrigation potential developed under different projects is about 103 m ha, which consists of 43 m ha under major and medium irrigation plan and the rest 61 m ha under MI plan. Eventually, there was a large gap in the utilization of created potential in irrigation program plan. In the Tenth plan and its last period, it was very clear that the total utilization of irrigation potential was to the extent of 88 m ha as against total created potential of 103 m ha, which depicted a gap of 16 m ha.

This program established a network of distributaries and minor over the command area. The projects initiated in the month of April 2004 were established with the objective of better water management practices and efficient utilization of irrigation water, which involves cost of water application. The work was already established in 0.9 m ha and was picked up by the end of the year 2004–2005 with the proper shape of the program.

The National Project for of Repair, Renovation, and Restoration of Water Bodies is directly linked to agriculture in the month of January 2005. The restoration program included physical progress work that has been completed for 733 water bodies. There was a MI scheme including groundwater and surface water schemes which were the major part of the plant for restoration and innovation. There were different sources of irrigation, MI has specific facilities as it is less capital intensive and requires less time to construct the plan in an efficient manner for future in which maximum utilization efficiency was the priority for the end users.

As the Indian rainfall pattern is very aberrant, Indian agriculture is still heavily rainfall-dependent on about 35% of the total arable area was irrigated, and distribution of irrigation across states was given proper dimension. Till the 11th Plan period, increased productivity, production, and ultimately the overall economy were observed.

4.15.1 IRRIGATION PLAN AND ITS DEVELOPMENT STAGE

4.15.1.1 STAGE 1: EXTENSION OF IRRIGATED AREA

With the view of the 5-year plans during 1950's, about 19% of the agricultural land was under irrigation system whereas 41% in Pakistan, 36% in Israel, 52% in Japan, and near 100% in Egypt.

4.15.1.2 STAGE 2: DEVELOPMENT OF MULTIPURPOSE PROJECTS

Challenge to the nation to meet up the need for food sufficiency. Irrigated state Punjab is at highest position about 73% followed by 50% coverage in Haryana state of the India. At the juncture, many of multi-purpose river projects were completed under the plans for extensive utilization.

4.15.1.3 STAGE 3: IRRIGATION COMMISSION AND ITS PRINCIPLES

The utilization of the irrigation water should be in efficient manner and increase in the water productivity is essential where sufficient irrigation facilities are not available for the judicious use of water. Optimum utilization of the irrigation water is needed where such facilities are not developed in abundance. Water conservation is the first priority to alleviate the water stress situation of crops.

4.15.1.4   STAGE 4: PROGRESS OF CUMULATIVE IRRIGATION POTENTIAL

The cumulative irrigation potential was achieved by 5th five year plan which was a considerable progress. Under the guidance and help of both Central and State governments, it was made possible. Table 4.5 shows the progress of cumulative potential and its utilization during different plan periods.

TABLE 4.5   Development of Cumulative Irrigation Potential and Its Utilization in Different Plan Periods.

S. no	Period	Cumulative potential	Cumulative utilization
1	Preplan	22.6	22.6
2	First plan	26.3	25.1
3	Second plan	29.0	27.7
4	Third plan	33.6	42.2
5	Fourth plan	44.2	48.5
6	Fifth plan	52.1	62.2
7	Sixth plan	67.9	68.6
8	Seventh plan	76.5	68.6
9	Eighth plan	89.5	80.7
10	Ninth plan	94.0	84.7
11	Tenth plan	102.8	87.2
12	Ultimate potential	139.89	-

4.15.1.5   STAGE 5: DEVE2LOPMENT OF AGENCY AND PROGRAM FOR BETTER UTILIZATION

During that time, the Government has set up NWDA for the utilization of water resources. The main aim of this project is to ensure early completion. There was an irrigation potential of about 8,503,000 hectares which was reported from major, medium, and MI projects. The CADP was with the Accelerated Irrigation Benefit Program (AIBP).

4.15.1.6   STAGE 6: IMPACT ON FOOD PRODUCTION

To increase the production of food grains with increasing irrigation area, the per capita per day food grains availability was also increased from 395 to ~512 g.

4.15.1.7 STAGE 7: RESTORATION OF WATER BODIES

There were enough feasible schemes for repair, renovation, and restoration of water bodies directly linked to agriculture. In India, the National Project funded was initiated for repairs, renovation, and restoration of water bodies. Under this scheme, restoration of water bodies has also been taken up with the help of World Bank.

4.15.1.8 STAGE 8: COMMAND AREA DEVELOPMENT PROGRAMME

During 1974–1975, the CADP was started for the maximum possible utilization of irrigation potential. The aim of the plan was to reduce the gap between the actual irrigation potential created and its utilization. These plans were also introduced for increasing the agricultural production from the irrigated commands. The network of distributaries was made up over the command area for the rotational supply of water and construction of field drains.

4.16 CONCLUSION

Effective harvesting of excess rainwater during heavy to medium rainfall period as well as artificial recharge of groundwater should fully be exploited wherever possible through participatory and mass awareness program to increase irrigation potential of the irrigation commands. Planned intervention is required to reduce the negative effects of surplus exploitation of groundwater by two-fold actions: (i) to control excessive draw down and (ii) to prevent water quality deterioration and degradation. The gap between technology generation and adoption in the farmers' field is to be minimized with the location-specific appropriate methods, systems, scheduling of irrigation based on different approaches, its transformation in farmers' language, water harvesting structures and water-lifting devices; generation of suitable technologies on rain water harvesting and recycling of excess water for its effective utilization for crop cultivation. Management of horticultural crops in orchards, flowers, and vegetable is recommended to find out suitable alternative of rice crop and to introduce less budget high-value crops. Development of submerged and mercy land (jheels and beels) having hydromorphic characteristics is also required for its proper utilization to increase agricultural productivity.

KEYWORDS

- **precipitation or rainwater**
- **groundwater**
- **surface water**
- **ocean water**
- **atmospheric water**

CHAPTER 5

Irrigation Projects in India

ABSTRACT

There are several observations regarding multipurpose projects having irrigation as a major component of the projects. There was an estimation of techno-economic expertise by the state design organization; the project status was under supervision by the state-level project appraisal/technical advisory committee that encompasses irrigation, agriculture, fisheries, forests, soil conservation, groundwater, revenue and finance departments, and state environmental management authority. In the light of these consideration, the techno-economic feasibility report is usually supported with an "Environmental Impact Assessment" and "Relief and Rehabilitation Plan" on environment and bio-safety. The project proposal was admitted to the Central Water Commission and shall be circulated among the members of the advisory committee of the Ministry of Water Resources for vivid scrutiny. Thus, in the process, the irrigation project is formulated to bring more and more land under irrigation potential. The proposed project if found acceptable will be recommended for investment clearance to the Planning Commission for inclusion in the 5-year plan or annual plan for implementation.

5.1 INTRODUCTION

In the present day's context, India needs to produce more food with limited land and water resources. The per capita water consumption is increasing and groundwater depletion is increasing. Water availability per capita was over 5000 m³/annum in 1950, nowadays, it is about 2000 m³. Wherein, the quality of available water is deteriorating day by day. Hence, there are lots of differences between basins and nonbasin zones.

Freshwater largely used in agriculture, accounting for about 80% of the water withdrawals for different purposes. It is very difficult to divert the water from agriculture to other sectors. It is very difficult to re-allocation of water out of agriculture as it can have a drastic impact on global food markets. It is estimated that the availability of water for agricultural use that may be reduced by 21% by 2020. Uses of every drop of water for efficient uses is done under irrigated farming.

With this view of the present and past water utility scenario, the policy reforms are very much needed. The new dimension of water utility should be present in judicious manner and it should also reform. Water use efficiency may increase with the introduction of appropriate water-saving technologies. Hence, it is needed that an integrated water use policy is formulated and judiciously implemented. Many international initiatives on integrated water use policy formulation and judicious implementation were taken in recent years.

5.2 IRRIGATION PROJECTS

In India, irrigation is the largest user of water ultimately it comes from precipitation. Many engineering structures were developed to collect, convey, and deliver water to areas. Irrigation projects covers small farm to millions of hectares agricultural land. Small irrigation project have low diversion weir, small ditches, and minor control structures (Johansson, 2005). Large irrigation project consists of a large storage watershed, a huge dam, networks of canals, branches, and distributaries and control structures.

5.3 CLASSIFICATION OF IRRIGATION PROJECTS

Irrigation projects are usually is classified as follows:

5.4 BASED ON CULTIVABLE COMMAND AREA

Major Irrigation Projects:

   I.   Major irrigation plan and its area coverage is more than 10,000 ha (CCA > 10,000 ha). Major irrigation plans cover huge watershed, water diversion structures, and network of canals. The aims of this project are multi-dimensional projects serving many aspects like flood control, tourist spot, hydropower, and recreation.

II. **Medium Irrigation Projects:** These projects cover a cultivable command area (CCA) less than 10,000 ha but more than 2000 ha. The objectives of this project meet up all the purposes; hence, these also fulfilled many aims. The main components of this type of project are medium-size storage, diversion, and distribution structures for supplying the irrigation water.

III. **Minor Irrigation Projects:** Minor irrigation (MI) projects consist of CCA less than or equal to 2000 ha are called as MI project. Under the MI plan, the water comes from tanks, small reservoirs, and groundwater pumping. These projects are run by different funding agencies and governing bodies. These are present individually within the command area.

MMI (major and medium) projects again classified into two, these are given below:

i.   **Direct Irrigation Method:**
     The direct irrigation method is age-old practice. The water is directly diverted to the canal from river. The canals are constructed by diversion structure like weir or barrage across the stream without storing water. Under this method, stream has an adequate perennial supply of water. This irrigation method is generally practiced in deltaic tracts and areas having level land.

ii. **Indirect or Storage Irrigation Method:**
     This method is very unique where the perennial flow of water less or negligible. In this method, rainwater is harvested and conserved in a reservoir during the monsoon. In this irrigation method, a dam is constructed across the river then the stored water is diverted to the crop fields through a network of canals. It is very useful for the lean period where no chances of precipitation. Generally, this method is adopted where the river is not perennial; such river is Mahanadi, Godavari, Krishna, and Kaveri.

Some irrigation schemes in India were classified by the planning commission and these are following:

i.   **Major irrigation schemes**: Under this scheme, CCA covers more than 10,000 ha.

ii. **Medium irrigation schemes:** Under this scheme, the coverage has a CCA of 2000–10,000 ha.

iii. **Minor schemes:** The minor scheme consisting of area coverage with CCA up to 2000 ha.

The cultural command area is the design on the basis of irrigation projects. The cultural command has been classified into two viz., a. Major and Medium irrigation Projects b. MI Projects:

Major irrigation projects are those projects which cost about Rs.5 crores and more; medium projects range between Rs. 20 lakhs and Rs. 5 crores and MI projects costs about Rs. 20 lakhs or less. On the other hand, during the year 1978–1979, Planning Commission has initiated a new classification for irrigation schemes, based on the areas:

1. Major irrigation area has more than 10,000 ha.
2. Medium irrigation schemes are those with CCA between 2000 and 10,000 ha.
3. MI projects having area up to 2000 ha.

Each of the two classifications is explained in subsequent paragraphs. Some points must be discussed here related to irrigation projects which may also be called irrigation schemes.

5.5   COMMANDED AREA

Commanded area (CA) is referred to as the area that can be irrigated by a canal network system, the CA may further be classified as under:

i.   **Gross command area:**
     Gross command area (GCA) refers the total area which is irrigated by a canal network system with unlimited quantity of water is available.

ii. **Cultivable command area:**
     CCA is nothing but the actually irrigated area within the GCA. However, the entire CCA is never put under cultivation during any crop season due to the following reasons:

     •   The land may be kept fallow that is without cultivation for one or more crop seasons to increase the fertility of the soil also. This is a cultural decision on the management of the particular areas.
     •   Some areas of the CCA irrigated water may not be applied as the crops get enough water from the saturation provide to the

surface water table due to high water table depth. During a crop season which is not cultivated is conversely termed as cultivable uncultivated area. Usually, the areas irrigated during each crop season are expressed as a percentage which represents the intensity of irrigation for the crop season.

5.6 MAJOR IRRIGATION PROJECTS OF INDIA

Several major irrigation projects are developed in India. Some of these are listed in Table 5.1

TABLE 5.1 Major Irrigation Projects.

Name of the projects	Name of river	Name of state	Cultivable command area (ha)	Year of completion
Bhakra Nangal Project	Sutlej	Punjab and Himachal Pradesh	4,000,000	1963
Beas Project	Beas	Punjab, Haryana, and Rajasthan	2,100,000	1974
Indira Gandhi Canal	Harike (Satlej and Beas)	Punjab	5 28,000	1965
Koshi Project	Kosi	Bihar and Nepal	848,000	1954
Hirakud Project	Mahanadi	Odisha	1,000,000	1957
Tungabhadra Project	Tungbhadra–Krishna	Andhra Pradesh–Karnataka	574,000	1953
NagarjunaSagar Project	Krishna	Andhra Pradesh	1,313,000	1960
Chambal Project	Chambal	Rajasthan and Madhya Pradesh	515,000	1960
Damodar Valley Project	Damodar	Jharkhand, West Bengal	823,700	1948
Gandak Project	Gandak	Bihar–Uttar Pradesh	1,651,700	1970
Kakrapar Project	Tapti	Gujarat	151,180	1954
Koyna Project	Koyna–Krishna	Maharashtra		1964
Malaprabha Project	Malaprabha	Karnataka	218,191	1972
Mayurakshi Project	Mayurakshi	West Bengal	240,000	1956
Kangsabati Project	Kangsabati and Kumari river	West Bengal	348,477	1956

5.7 DIFFERENT IRRIGATION PROJECTS: POTENTIALLY CREATED AND UTILIZED

Demand for irrigation water in the country is huge. As per the limitations, there are limits in storage structure where water restricts the potential for irrigation. The main aim of the water basin is to transfer water and also helps in recharging groundwater. Potential Created (PC) and Potential Utilized (PU) are presented in Table 5.2.

TABLE 5.2 Project Wise UIP, PC, and PU Till End of IXth Plan (in Million ha).

Irrigation projects	UIP	PC	PU
Major and minor irrigation project	58.47	37.05	31.01
Minor irrigation (MI)			
Surface water	17.38	13.6	11.44
Groundwater	64.05	43.3	38.55
Sub-total	81.43	56.9	49.99
Total	139.9	93.95	81.00

PC, potential created; PU, potential utilized; UIP, ultimate irrigation potential; MI, minor irrigation

5.8 DEVELOPMENT OF MAJOR AND MINOR IRRIGATION PROJECTS IN INDIA

The new irrigation project was aimed to make a report based on proposal and rules for submission, appraisal, and clearance of Irrigation. As per the result, the project appraisal certifying the following points:

1. Entire proposal for planning of irrigation project and economic feasibility have been carried out depending on certain proposal.
2. About 10% of the command area has been investigated in three clusters of land representing terrain conditions in the command for calculation of the conveyance system.
3. Around 10% of the canal structures have been investigated in deep.
4. Minute sub-surface, surface construction material investigations have been carried out. Out of the work plan, there are dams, weirs, canal, and distributaries of about 10 meter per second.
5. The required designs for the various elements of the project have been done as per the guidelines and relevant Indian standards. Required investigations have been done with special regard to the problem of water stagnation. Most of the plans have been made for conjunctive use of groundwater and its drainage.

5.9 ENVIRONMENTAL IMPACT OF IRRIGATION PROJECTS IN INDIA

Many water projects have many objectives, such as irrigation, hydro-electric power, flood water supply, and ecosystems and environment for agricultural sustainability. Hence, the projects are properly formulated and suitably designed. Some adverse effect is environmental evaluation and assessment are complete at the planning and design stages of the concerned project. Thus, this is an urgent need follow-up the impacts on the environmental assessment system to quantify the impacts of irrigation and drainage projects.

The objectives of the assessment are the process of identifying, predicting, evaluating, and mitigating the biophysical, social, and other related impacts of development proposals. Environmental impact assessments were special in which they do not require close to predetermined environmental results. The importance of the environment has a key role in their decisions and to evaluate those decisions in view of detailed environmental research, and public comments on the efficient environmental impacts of the proposal are to be evaluated.

5.10 MICROIRRIGATION PROJECTS

Microirrigation projects are under the Pradhan Mantri Krishi Sinchayee Yojana (PMKSY). This is drought-proof and producing "more crop per drop." Many trials of the projects such as Pradhan Mantri Krishi Sinchayee Yojana are aimed at investment in irrigation and improving the efficiency of water use in agriculture area.

Irrigation in the Indian context is about 48.7% of the total sown area means two-third of the population involved. The area of the microirrigation system stands top of its target for 2015–2016. During 2016–2017, the 839,000 ha area was achieved which was more than the target. The part of the data for 2017–2018 reflected the government's success regarding improvement. The target was given during 2017–2018 which also fulfilled 1.2 million ha, and partial data from states evaluated that 926,432 ha had been covered wherein the achievement is likely to enhance as compilation and presenting of works undertaken in the financial year have to be updated.

Most of the northeastern states have made zero improvement regarding the irrigation projects. Uneven progress means meeting long-term goals that can be challenging to fulfill within a scheduled time. There were five states Andhra Pradesh, Karnataka, Gujarat, Maharashtra, and Tamil Nadu account for 78% covered. It was found unique observation, among the laggards;

Bihar was able to implement only 86 ha while Himachal Pradesh added 1107 ha. In all respect, it was noticed that lead performer Andhra expanded microirrigation coverage about 186,444 ha. Karnataka also covered about 164,967 ha in microirrigation. The position of the states like Gujarat ranked third and covered about 143,134 ha under the irrigation network. In centrally sponsored schemes like Pradhan Mantri Krishi Sinchayee Yojana, the Centre shares about 60% of funds and states funded 40% of total of 100%. There was a major reason for the unequal distribution of central schemes that some states are not able to allocate their 40% share in state budgets allocated. India's net irrigated area is 68.38 million ha, out of it about 71.74 million ha are un-irrigated. Microirrigation and its components were implemented and farmers get benefitted toward livelihood security and family farming.

5.11   CONCLUSION

It is to be concluded that judicious use of water with an integrated manner in agriculture system is highly required. There is an urgent need of assessment for the various issues, regulatory concerns, water laws and legislations, research and technology development and dissemination, social mobilization, and participatory and community involvement. In addition to this gender and equity concerns, economic evaluation is more important for efficient water use. The future water research institution should have its function in a trusteeship mode for sustainable water security and efficient uses.

KEYWORDS

- **precipitation or rainwater**
- **groundwater**
- **surface water**
- **ocean water**
- **atmospheric water**

CHAPTER 6

Role of Water in Plant System

ABSTRACT

The state of water supply to the plant in a system is governed by soil moisture regimes. Low evaporative demand along with unlimited soil moisture supply controls ET (evapo-transpiration) at its potential rate. High evaporative demand and low moisture supply, the control to the potential rate is comparatively much less. Under intermediate conditions, the control may be either from soil, the plant, or the atmosphere. The plant supports root penetration, drainage, aeration, retention of moisture, and plant nutrients. These are linked with the physical condition of the soil. Some important physical properties of soils are soil texture, structure, density, porosity, consistence, and soil water. Water helps in nutrient availability to the plant, helps to be absorbed through roots used for cooling, helps in convertion to vapor and escapes into atmosphere through plant leaves; the process is called transpiration. Water can also escape directly either from soil surface or open water surface into atmosphere; the process is called evaporation which combinedly termed as evapo-transpiration (ET) that is actual term of water use in plant system as well as the basis of water productivity and water use efficiency.

6.1 INTRODUCTION

Water is an abundant source covering about 71% of the total surface water and uneven distribution of water hampers crop production. Due to the less availability of water, there are significantly larger losses in production and vegetative cover in the terrestrial ecosystem. The physiological importance of water in plants is essential for successful plant growth, involving photo-synthesis and several other biochemical processes. It characterizes the growth and productive behavior of plant species. It has a relationship in crop water use with scarce water sources. The physiology of water inside the plant plays a role in the absorption and loss of water from the soil as well as plant parts. The water absorption also depends on the crop variety and species. The transpiration process of plants transmits water to the atmosphere and very less percentage is utilized by the plants for their biochemical activities.

Irrigation is the artificial application of water to arable lands in order to supply crops with water requirements that are not satisfied by natural precipitation. Adequate food and fibers cannot be produced without irrigation water under dry conditions. Uneven distribution and erratic rainfall patterns badly affect the crop yield where in humid and sub-humid regions, yield loss may not occur with an exception. Crop water loss occurs when evapotranspiration rate is reached at its peak point. Different soil properties also affect the soil water loss through evapotranspiration.

6.1.1 PLANT AVAILABLE WATER

Generally, it is the amount of water held by the soil between field capacity (FC) and permanent wilting point (PWP). In view of this definition, FC is the amount of water in a well-drained soil. For coarse-textured soils, drain off occurs soon after providing water to the field. Coarse-textured soil having large pore space results in high water infiltration. Heady soil having fine-textured soil and it has high water holding capacity. FC refers to when the field is full of water just after irrigation whereas PWP refers to a state at which plants can no longer obtain enough water. Subsequently, plant starts to wilt and after providing water it will not revive. Plant water availability varies between FC and PWP. Under this range, metric water potential lies in soil water system. Fine-textured soil has greater the available water with no much soil salt.

6.1.2 PLANT ROOT SYSTEMS AND SOIL WATER

The plants absorb soil water and nutrients through root systems, that is, fibrous roots and taproots. Graminaceous or Poaceae family has monocotyledon and fibrous root systems. Monocot crops are rice, wheat, etc. Most of the dicoty-ledonous plants such as peas, lentils, sugar beets, and alfalfa, have taproot systems. Radical is the first root appearing from a seed, is a seminal root. The other part grows above the ground is plumule. From the primary root, secondary and lateral branches come. In maize plant, roots may also develop from above ground nodes such as the brace roots. Tap root penetrates much deeper as compared with the fibrous root system. The tap root having less root branch is compared with tap root system. The main aim of this type of root is to better anchorage. The monocotyledons and dicotyledons plants differ with clear root length and complex branching root system. The growth pattern of primary branch is more as compared with the secondary branch followed by the lateral roots. Root elongation is observed as high as 2.4 inches per day in maize. Under unfavorable conditions, root growth may severely be restricted.

6.1.3 PLANT ROOT DEPTH AND SOIL WATER

The roots depth gives anchorage and uptake nutrient from the soil. The vegetative stage of plants with only shallow roots will not uptake soil water from the deeper layer of soil. Plants water extraction pattern is unique and about 40% of their water needs from the upper layer of root zone, then 30% from the next layer, 20% from the third layer, and takes only 10% from the deepest part of the soil. Hence, plants will extract about 70% of water from the top half of their total root penetration. It reveals the depth of root penetration and 70% water extraction for many field crops. Lower layer of the root zone reserves higher percentage of the crop's water needs. It depends on the crop types and its root proliferation to browse soil water. Supplying water to deeper depths subjects the irrigation to a higher potential for deep percolation losses and aims to maintain optimum water use for the maximum water use efficiency. Clay and clay loam soil are free to restriction whereas in sand and gravelly soil. There is restricted root penetration (Table 6.1).

TABLE 6.1 Depth of Root Penetration and 70% of Their Water Extraction for Several Common Field Crops.

Crops	Depth of root penetration (ft)	70% of their water extraction (ft)
Corn	4–6	2–3
Sorghum	4.5–6	2–3
Alfalfa	6–10	3–4
Soybean	5–6	2–3
Wheat	4–6	3
Sugar beat	5–6	3

6.1.4 CROP SPECIES AND ROOTING CHARACTERISTICS

Optimum irrigation provides quality rooting and rooting density. In crop, variety to variety the root behavior is different. Genetically, characteristics consist of diverse rooting pattern. Different types of crops have different plant roots browsing capacities. The crop root systems are badly affected due to soil types. Coarse soil, stony soil, gravelly soil hamper root penetration and nutrient uptake. Only balanced nutrient can be taken from fertile soil with high organic, heavy soil with good aeration. Water uptake pattern is depending on the root behavior, and soil compaction reduces root growth rates with total root exclusion. Hardiness of soil reduces the volume due to mechanical impedance to root expansion which lowers the rate of gas exchange between the soil and atmosphere. The root growth system provides a negative tension when extracting water from the ground level.

The water potential between soil and plant system, the plant must provide at least 0.3 bars of negative tension. The maximum negative tension provides balanced soil water tension. Wilting point of plant refers no longer plants take water from the soil and it is daily plant water need as influenced by climatic parameters and root and water availability under the soil. Water is needed at different stages of growth, that is, at vegetative stage of the plant; this stage requires less water than reproductive stage. During the maturity stage of the plant, water requirement is very less or negligible.

6.1.5 PHYSICAL PROPERTIES OF SOIL WATER

Soil water relations are interrelated with physical properties of soils. Soil water affect movement, retention, and absorption of water by plants. Easiness of plant water and nutrients uptake improves an irrigation system.

6.1.5.1 SOIL PROFILE SYSTEM AND COMPOSITION

The soil has wide diversity and it composes viz, solid, liquid, and gaseous, particulate, disperse and porous system. Two states of soil, that is, solid phase constitutes soil water with dissolved substances. The soil also contains soil solution and the gaseous phase part. The soil profiles have solid soil matrix including chemical and mineralogical composition. The soil particles have different in size, shape, and orientation. Amorphous substance of soil contains organic matter later on it mineralized.

The different soil layer under varying situation, that is, dry season or rainy season, its aggregation also somewhat changes. Soil aggregate determines the geometric characteristics of the soil water. Successful crop production depends upon the soil type and water availability. Soil profile is a reservoir for moisture reservoir as per the need of plants and it also depends on the crop growth stage.

6.1.5.2 SOIL PROPERTIES AND SOIL-WATER-PLANT RELATIONSHIP

a. **Soil Depth**
   Arable soil layer where rhizosphere exists is referred as soil depth. Soil depth also lies over hard rock or hard bed rock layer below which roots cannot penetrate. The soil depth is directly related to the rhizosphere. The rhizosphere contains water storage capacity, nutrient supply, and feasibility for land leveling and land shaping.

   Light soil has low moisture-holding capacity and nutrient is leached through water and irrigation is needed at regular intervals. The shallowness of soil is further unfavorable in areas needing land leveling and shaping. Uneven distribution of land affects soil water relations besides nutrient retention capacity and nutrient availability. Heavy soil has a high water holding capacity and better root growth.

b. **Soil Texture**
   Soil texture is the most vital property of the soil. Soil texture is intimately related to soil water and plant growth relationship. Soil texture has mineral particles of various sizes. There are three particle sizes, such as coarse, medium, and fine particles. Generally, these particles are named as sand, silt, and clay. There are three soils textural classes which are known as sandy, loamy, and clayey.

   The soil texture has the following properties:

   a. Water holding capacity
   b. Irrigation water depth
   c. Frequency of irrigation
   d. Irrigation interval
   e. Movement of water and air
   f. Infiltration rate

Sandy soils have the low water holding capacity with high infiltration rate. It also facilitates high air movement. Sandy soil requires frequent irrigation with high water depth. Besides, clayey soil has relatively high water-holding capacity. Moreover, this soil faces water logged during high rainfall due to lack of poor infiltration rate. Considering its various effects, the soils with loamy texture are the best soils for growing all types of crops under irrigated conditions.

c. **Soil Structure**
   The structure of a soil indicates the arrangement of the soil particles. Aggregation forms after the adhesion of smaller particles. The soil structure is related to soil tilth, where good tilth provides better soil health and soil environment.

The different shapes of aggregates depend on the structure of the soil. The soil structures are of the following types:

a. Spherical, for example, granular or crumbly.
b. Platy, for example, columnar or prismatic subtypes.
c. Blocky, for example, cube and sub-angular.

The influence of soil structure

1. Air and water permeability
2. Total porosity and water storage capacity
3. Penetration and proliferation of roots

Clay particles have wide diversities in sizes namely sandy and clay where sandy soils allow water to percolate either too rapidly or too slowly. Platy structure of soil obstructs the downward movement of water. Among the soil structures crumbly, granular, and prismatic structure are efficient for irrigation management and better plant growth.

d. **Soil Density**
   There are two types soil density.

i. **Particle Density**
   Particle density (PD) is the ratio of a given weight of soil solids to the volume. PD is expressed in g/cm^3. Soil minerals have wide range of density and the values for most mineral soils usually vary between 2.6 and 2.75 g/cm^3. This property of soil is largely

independent of size. The PD is always required for the estimation of total, capillary, and noncapillary porosity.

ii. Bulk Density

Bulk density (BD) defines the ratio of a given mass of an oven-dried soil to volume. So, BD is the total mass of soil particles, that is, solids + pore spaces. Soil having more numbers of pore space in turn the high soil water storage capacity. This phenomenon affects the crop performance particularly where water availability is limited.

iii. Adhesion and Cohesion

Two basic forces responsible for the soil water retention and water movement is given below:

a. Adhesion

It is the attraction between water molecules and solid surfaces, that is, soil minerals.

b. Cohesion

Cohesion is nothing but attraction of water molecules for each other. With the help of adhesion, water molecules are present rigidly at the surfaces of soil solid. Soil particles are tightly bound with water molecules that are hold by cohesion with other water molecules. The adhesive force diminishes rapidly with distance from the solid surface. The cohesion of one water molecule to another and water molecules are changing in size and shape. The attraction of water molecules for each other also facilitates the solid. Hence, adhesion and cohesion force is helpful for the soil solids to hold water.

c. Moisture Tension

The moisture occurs in the soil against gravity and it also refers to moisture tension. The soil moisture tension is a measure of the tenacity with water. It is retained in the soil and reflects the force per unit area shown by plants to remove water from the soil. Several units have been used to express the force with which water is held in the soil. Soil water tension is measured in bar which equals to the pressure exerted by a vertical height of water column.

Soil water tension is instantly measured, a pressure of one bar is approximately equal to the hydrostatic pressure exerted by a vertical column of water. The water column heights have a height of 1023 cm. As we know 1.0 bar is equal to 0.9869 atmospheres. This value is the average air pressure at sea level, that is, equal to 14.7 lbs/in^2. The suction of water having a height of 10 cm is equal to 0.01 bars or 100 milli bars. Where, 1.0 bar is equal to100 centi bars.

d. Water Movements in Soil

i. Infiltration

The entry of water from one layer to another is termed as infiltration. It is the process of water entry in to the soil; basically by downward flow through the soil surface is termed as infiltration. Infiltrability refers to the volume of water flowing into the soil surface area per unit time and infiltrability is the similar term of infiltration.

ii. Cumulative Infiltration and Rate of Infiltration

The infiltration rate is not constant over time due to several reasons. Infiltration rate is decreased when soil gets dry. This approach constant rate is often termed as basic intake rate or steady-state infiltration rate. The cumulative infiltration, being the time integral of the infiltration rate, has curvilinear time dependence. Generally, infiltrometer is used for measuring the infiltration rate of a soil. The variation of infiltration rate is different class of soil. The constant rate of infiltration rate for various soil types is influencing infiltration rate and is characterized by time from the onset of rain or irrigation.

i. Seepage

The seepage process occurrs on the soil surface. In this process, horizontal movement of water through soil pores is taken place in the soil profile under unsaturated condition is known as seepage.

ii. Permeability

Permeability process meant the penetration of water and air through the soil pores. Permeability of soils is generally categorized as rapid, moderate, and slow. The rate of permeability differs with varying soil texture. It is rapid in coarse-textured soil sands and slow in heavy-textured soils.

iii. Deep Percolation

Deep percolation refers to infiltration process when it is a transitional phenomenon that occurs at the soil surface. The water moves downward into the soil layers. After infiltration, water moves downward within the soil profile under the influence of both gravity and hydrostatic pressure and is termed as deep percolation. Sandy soils facilitate higher percolation rate as compared with clayey soils. Loss of water by percolation in agriculture fields is generally less compared with the uncultivated soils.

iv. Hydraulic Conductivity

It is a measure of the soil stability to transmit water when submitted to a hydraulic gradient. It is defined by Darcy's law, which is one-dimensional at vertical flow. The velocity refers to as per laws the average velocity of the soil fluid through a cross-sectional part in the soil where flow takes place. The cross sectional part is height, that is, h^1 and h^2 are hydraulic heads, and L is the vertical distance. The coefficient of proportionality is K. This K is called the hydraulic conductivity.

6.1.6 TYPES OF WATER MOVEMENT

Water movements in the soils are multidirectional because of various states and directions in which water moves. Water is dynamic in soil component; generally three types of water movements are known, that is, saturated flow, unsaturated flow, and gaseous movement. Two flows involve liquid water in contrast to gaseous flow. Water flows in effect to energy gradients, with water moving from a zone of higher to one of lower water potential.

a. Saturated Water Movement

When all the macro and micropores are filled with water, the soil is said to be at saturation. Water flow in macro or micropores under this soil situation is known as saturated flow. The saturated flow is estimated by two factors, that is, the hydraulic force and soil pores that allow water movement.

The quantity of water per unit of time K that moves through,

A   = the cross-sectional area of the column,

Ksat = the saturated hydraulic conductivity,

$\Delta\Psi$ = the change in the water potential

$\Delta\Psi/L$ = water potential gradient

Ksat unit is cm/s or cm/h.

b. Unsaturated Water Movement

This is the situation when the soil macropores are almost filled with air and the micropores with water and some air; and under such conditions any water movement that occurred under this soil condition is termed as unsaturated flow.

Water movement under field conditions, most of the soil pores are not completely saturated with water. Under this situation, it is very sluggish compared with that when the soil is at saturation. This is because at or near zero tension, the tension at which saturated flow occurs. It lies at tensions of 0.1 bars and above, which characterize unsaturated flow.

At low tensions unsaturated soil conditions, the hydraulic conductivity is higher in sandy soils, whereas in clay soil, it is lower. Unsaturated flow affected by flow of water, that is, its direction and rate and this gradient is the difference in tension of both capillary ends. The water moves from lower tension to higher tension. The force responsible for this tension is the attraction of soil solids for water. The higher the water status leads to higher moisture tension and rapid water movement.

6.1.7 *PHYSICAL WATER*

The water held within the soil pores space is termed as to as soil moisture. Physical water is held in the soil and it flows into plant system. It depends on dry and wet soil. Soil water may be classified into three types' viz, gravitational water, capillary water, and hygroscopic water.

a. **Gravitational Water**

Water held between 0.0 and 0.33 bars soil moisture tension. Gravitational water is free and it not used by the plants. Plant cannot uptake this water. It moves rapidly downward to toward the water table under the influence of gravity and is known as gravitational water. Gravitational water is held with low tension, it is of little use to plants. It occurs in the larger pores, that is, macrospores, lead to low soil aeration. Natural drainage is required for the ideal plant growth.

b. **Capillary Water**

This water is held in the capillary tube. Capillary water is held in micropores around the soil particles by the characteristics of adhesion, cohesion, and surface tension. Capillary water is held between FC and hygroscopic coefficient. The water within the capillary range is not equally available, that is, it is easily available from 0.33 to 15 bars or critical soil moisture level. Unavailable water lies between 15 and 31 bars. Unavailable water is held very tightly in thin films and is practically not available for plant use but it can take only some soil microorganisms. This moves in all directions but always in the direction of increasing tension.

c. **Hygroscopic Water**

Hygroscopic water is present tightly in thin-film layer of 4–5 mili micron thickness and it is present on the surface of soil colloidal particles at 31 bars tension and above is referred to as hygroscopic water. It moves in gaseous form and plants cannot absorb such water because it is held in vapor form by the soil particles at a tension of more than 31 bars. This water can be utilized by few soil microorganisms. Hygroscopic water easily evaporates at atmospheric temperature. Hygroscopic water is difficult to separate from the soil even though it is heated at 100°C and above for 24 h or more.

6.1.8 SOIL MOISTURE CONSTANTS

The water contents expressed under specific standard conditions are termed as soil moisture constants. They are used as reference points for practical under the irrigation water management. The energy status of soil water for soil water constants gives useful knowledge for understanding and interpretation.

6.1.9 *SATURATION CAPACITY*

Saturation capacity defines the condition of soil at which all the macro- and micropores are filled with water. Saturation capacity of the soil is at maximum water retention capacity. Under the saturation capacity, the metric suction at this condition is essentially zero as the water is in equilibrium with free water. Above saturation capacity, the excess water of soil is lost from the rhizosphere as free water.

6.1.10 *FIELD CAPACITY*

Field capacity refers to the amount of water present in the soil after excess water has been drained out. The rate of downward movement has decreased, which usually takes place within 1–3 days after a rain or artificial application of water under the same pervious soils having uniform texture and structure. The soil moisture tension at FC varies on the soil texture range from 0.10 to 0.33 bars. It is assumed that as the upper limit of available soil water. The FC is greatly influenced by the size of the soil particles, finer the soil particles higher the water retention due to very large surface area and vice versa.

6.1.11 *PERMANENT WILTING POINT*

Permanent wilting point of soil is referred as the water that is held so tightly by the soil particles that the plant roots can no longer receive sufficient water. The PWP and nearly all the plants growing on such soil show wilting symptoms. Once this plants is facing this stage and plant do not revive in a dark humid chamber unless water is supplied from an external source. Under this moisture tension, PWP is reached at 15 bars. PWP is referred to as determined by growing indicator plants such as sunflower in small containers. To determine the moisture content by the pressure membrane apparatus this can be required at 15 bars.

6.1.12   AVAILABLE WATER

The amount of soil moisture held between two cardinal points viz., FC and PWP as available soil moisture.

The soil moisture is held below the PWP, it is held so tightly by the soil particles as a gaseous form that the plant roots are unable to uptake it rapidly enough to prevent wilting. Thus, this water is not useful for the plants and forms the lower limit of available soil moisture.

6.1.13   HYGROSCOPIC COEFFICIENT

Hygroscopic water is the amount of water that is held when it is in equilibrium with air at standard atmosphere, that is, 98% relative humidity and at room temperature. In other words, it is the amount of moisture absorbed by a dry soil at any given temperature, expressed in terms of percentage on an oven dry basis. The matrix suction of soil water lath is moisture content that is nearly about 31 bars.

6.2   CONCLUSION

The demands for water to the crop usually comes from evaporation and transpiration which is commonly known as consumptive use associated with unavoidable conveyance losses during application and some special needs during land preparation and transplantation. The supply of water to meet the crop needs usually comes from effective rainfall, irrigation water application, and contribution from storage moisture at the soil profiles. So to calculate the water requirement of the crop, it is essential to know the existing moisture of the soil on which the crop is growing. Knowledge of rooting habits of a particular crop and procedure to determine the moisture present in the different soil profiles up to root zone depth are prime importance. The crops usually extract moisture for essential activities through root system. Water comes up to the tip of the leaves to complete the food production processes and goes down up to roots again. Therefore, it is required to know the crop(s) to be grown including its rooting habits, physiological critical growth stages of the crop and the existing moisture status soil profiles up to root zone depth varies crop to crop.

KEYWORDS

- soil depth
- soil structure
- soil texture
- soil density
- adhesion
- cohesion
- moisture tension

CHAPTER 7

Water Productivity in Agriculture

ABSTRACT

The term "water productivity" is introduced to complement existing measures of the performance of irrigation systems, mainly the model irrigation and effective efficiency. Typical irrigation efficiency focuses on establishing the nature and extent of water losses and included storage efficiency, conveyance efficiency, distribution efficiency, and application efficiency. These measures are particularly useful for water system managers who use them to (a) assess how much water they were losing in the storage, conveyance, distribution, and application subsystems and (b) identify interventions for improvement of the performance. Water saving is the process of reducing nonbeneficial water uses and making the water saved and available for a more productive use. In the situations where water is scarce, reducing nonbeneficial uses becomes one of the main ways for reducing water scarcity. Improving water productivity seeks to get the highest benefits from water and hence can be viewed as a major contributor to water resource development.

7.1 INTRODUCTION

Despite the fact that more than two-third of the Earth's surface is covered with water, currently, near about 450 million people in 29 countries are facing severe water shortage and at least 20% more water would be required to feed additional 3 billion populations by 2025. About 97.5% of ocean-sea water is not available due to its salinity and 2.49% is locked up in ice and only 0.01% is technologically available and economically utilizable water either from surface or groundwater sources for human uses. Hence, water-saving technologies developed so far is very pertinent to propagate amongst the farmers. Nowadays, water is becoming a serious and scarce commodity to maintain the civilization amidst of plenty. Water resource management is a multidisciplinary activity; managing irrigation water needs agronomy and crop husbandry; efficient methods and system of irrigation needs soils scientists and engineers. Optimum use of water at the right time is the key to successful management of this precious resource. Above 98% of the irrigated field is under surface irrigation where more than 50% of water is assumed as wastages of water in different ways. The drip and sprinkler irrigation methods have become the tool of efficient water use.

7.2   WATER PRODUCTIVITY AND EFFICIENCY

Water productivity (WP) is a measure of performance of water expressed as the ratio of output to input. WP may be calculated and it could be for one of the inputs of the production system in agriculture.

7.3   TOTAL PRODUCTIVITY

Total productivity is the ratio of total outputs divided by total inputs. It depends on partial or single-factor productivity. Total productivity refers to the ratio of total tangible output to input of one factor within a system of management. The factors affecting total productivity is water, land, capital, labor, and nutrients.

7.4   WATER PRODUCTIVITY (WP)

WP refers to as like land productivity. It is a partial factor productivity that measures how the systems convert water into agricultural produce. The following equation is stated below:

Water productivity (WP) = Output comes from water use

WP measures the performance of irrigation systems. It measures classic irrigation and effective efficiency. Classic irrigation efficiency (IE) describes on establishing the nature and extent of water uses. IE also focuses on storage efficiency, conveyance efficiency, distribution efficiency, and application efficiency. These measures are useful for water users in the following ways:

a. To assess how much water losses in the storage, conveyance, distribution, and application.
b. To identify interventions to improve performance of use of water.

It is required to know the performance of water use in a large system. In some cases, that is, basin or subbasin, unable to know use of water catchment area. Basin irrigation system is big way and its water losses assume reuse in the basin area, that is, deep percolation and runoff losses.

The concept of effective efficiency, which takes into account the quantity of the water delivered from and returned to a basin's water supply. In an irrigation context, effective efficiency is the amount of beneficially used water divided by the amount of water used during the combined processes of conveying and applying that water.

The WP is a holistic and integrated performance assessment by:

i. Incorporation of all types of water uses in a system.
ii. Inclusion of all types of outputs.
iii. Integration in technical and allocation efficiency.
iv. Incorporation multipurpose basin.
v. Inclusion of multiple sources of water.
vi. Integration of non-water factors that affect productivity.

7.5 WATER PRODUCTIVITY AND WATER SAVING

Water saving is referred as the process of making the water available for a more productive use. Under water stress condition, reducing non-beneficial uses guide reducing water scarcity. Improving WP has benefits from water and that can be a major contributor to water saving.

Water saving technology adoption reduces non-beneficial depletion that can be accomplished through the following points:

i. Minimization of runoff and
ii. Minimization of evaporation.

In view of improving, IE is considered to be the most minimization of evaporation loss. Non-beneficial water use and its reuse, for example, runoff losses may be the main losses. Water saving has a negative impact on water quality, drinking water supply, and groundwater balance.

Water conservation focuses on rain water harvesting in a bigger way. Increasing WP needs other resources with high priority, such as labor, capital, and its management.

Basin and catchment area management has the following points:

i.   To minimize runoff and deep percolation.
ii. To reduce non-beneficial use of water.
iii. To save water through reduced runoff and deep percolation losses.

7.6   FREQUENTLY USED TERMS IN WATER PRODUCTIVITY

i.   **Gross inflow**

   Gross flow is the total amount of water flow from precipitation, rivers, and groundwater.

ii. **Net inflow**

   Net flow is any increase in storage in the surface soil or groundwater.

iii. **Available water**

   Available water refers as net inflow minus the committed and nonutilizable outflow.

iv. **Depletion of water**

   Water depletion refers to utilization of water within the system.

v.   **Process depletion**

   It is that amount of water diverted for use that is depleted by users.

vi. **Nonprocess depletion**

   Nonprocess depletion represents when water is depleted by other means, but not by a human intended process.

vii. **Nonbeneficial depletion**

   It refers when water is depleted through evapotranspiration.

viii. **Committed outflow**

   It is that part of outflow from the given area that is committed to other uses.

ix.   **Uncommitted outflow**

Uncommitted outflow refers to part of outflow that is not committed. Many times it is of the uncommitted outflow can be nonutilizable.

x.   **Conventional irrigation efficiency (CIE)**

CIE is measured with conventional or traditional methods. The IE can be measured by using IE such as conveyance efficiency, distribution efficiency, and water application efficiency.

xi.   **Duration of water in fields (DWF)**

DWF is the total number of days water is maintained in paddy fields from transplanting until harvesting.

xii.   **Field operation and management efficiency (FOME)**

FOME is the operation or management of irrigation water in the field.

xiii.   **Integrated water resource management (IWRM)**

IWRM is the management of water resources to all users and beneficiaries.

xiv.   **Irrigation efficiency (IE)**

IE is the ratio of the amount of water uptake by the plant to the amount of water supplied through irrigation by means of surface, sprinkler, or drip irrigation.

xv.   **Irrigation situation efficiency (ISE)**

ISE is the efficiency of an irrigation system assessed at special time that takes into account the micro effectiveness of irrigation water use.

xvi.   **Nested system efficiency (NSE)**

NSE is the efficiency of two or more connected system that allow single source of water supply.

xvii.   **Net water requirement (NWR)**

It is the amount of water required to meet up evapotranspiration and deep percolation or root zone soil moisture deficiency.

xviii. **Gross water requirement (GWR)**

GWR is the actual amount of water applied to meet crop evapo-transpiration and or percolation observed under field conditions.

xix. **Nested system productivity indicator (NSPI)**

It is the indicator of WP for the two or more connected systems on the same time of water use.

xx. **Nested system wetting days (NSWD)**

NSWD is used all time for wetting the soil for land preparation (puddling) especially for paddy crop for connected systems of irrigation water use.

xxi. **Productivity of Water (PW)**

PW is the ratio of output, that is, physical, economic, or social to the amount of water depleted (irrigation water applied) in producing the output.

xxii. **Nested System Productivity (NSP)**

NSP is the PW for a nested system (water channel connected with many ways) of water use.

xxiii. **Relative Nested System Productivity (RNSP)**

RNSP is WP in two or more nested water systems (multiple irrigation channel connection).

xxiv. **Standard Rice Water Productivity (SRWP)**

SRWP is the ratio of original rice yield harvested under the net annual paddy water requirement.

xxv. **Water Depth Efficiency (WDE)**

WDE is the efficiency and it is the ratio of required to actual water depths held in paddy fields during active vegetative stage.

xxvi. **Water Use Efficiency (WUE)**

WUE is defined as the ratio between the amount of water used by users and the total amount of water applied for field.

xxvii. **Whole System Efficiency (WSE)**

WSE is the efficiency of water use for the entire well-structured irrigation system.

xxviii. **Whole System Productivity Indicator (WSPI)**

WSPI is the WP indicator for the entire well-structured system of water use.

7.7 WATER PRODUCTIVITY ASSESSMENT APPROACHES

i. **Water input assessment**
Water can be assessed before the water used in a production system. The water distribution system exhibits its pathways, quantifies inflows, depletions, and outflows. Utilization of water while entering and draining is the key role in irrigation system. Water while resources availability includes precipitation, surface, and groundwater. Water losses within the system occur mainly in two ways; these are depletion and outflow.

The supply of water in production can be depicted in many ways depending on the system boundaries and the level of detail gross inflow to a field. These are following:

i. Net inflow of water.
ii. Available water.
iii. Both beneficial and nonbeneficial depletion of water.
iv. Beneficially depleted water.
v. Process depleted water.

ii. **Assessment of outputs**
This is an advantageous of outputs which is derived from using water. Output assessment can be quantifiable or nonquantifiable. Output assessment can be generated either through depleting or nondepleting uses of water. Output assessment provides a wide range of advantages, with quantifiable and nonquantifiable sector.

iii. **Water productivity indicators**
WP can be applied at different ranges to adopt the needs of different stakeholders. This is gained by defining the inputs and outputs of water in units suitable to the users need.

The output is referred in the following points:

i. Physical output refers to biological yield.
ii. Economic output is economic yield, it refers to either net return or gross return.
iii. Harvest index also refers to biological output divided by economic output. It can also be expressed as percentage.

The unit of water stands as volume (m³) as the highest opportunity cost in alternative uses of the water. For assessing WP in a different cropping system, different types of scales are given below:

a. **Crop scale**
Crop scale is to assess how efficient a particular crop or cultivar of a crop is converting water into biomass or biological yield. This scale can be quantified by with the output either as biological yield or crop yield or harvestable produce.

b. **Field sale**
Field scale is to assess the beneficial uses. A particular cropping system converts efficiently water into beneficial output, that is, food grain, pulses, oil seeds, vegetables, and others. It can be quantified as total biomass or crop yield and the water inputs are the amount of water used.

c. **Field farm scale**
It is used for beneficial and as well as non-beneficial purpose to assess the opportunities of saving water loss. It can be quantified as biological produce, crop yield (kg), and crop value (Rs). The water depletion takes place from the system through (a) evaporation (b) flows to sinks, (c) water pollution, and (d) use into the product.

d. **Irrigation system scale**
The irrigation system manager expresses assessing how productively the water presents to the irrigation system. The amount of water depleted and recaptured for recycling. It can be quantified in physical and economic term.

e. **Subbasin scale**
Subbasin scale is an assessing option for increasing WP. The output can be quantified as either biomass or difficult to regulate subbasin scale assessing the opportunities for involving in water infrastructure.

f. **Basin scale**
    Basin scale is assessing the productivity of the renewable water that
    enters into the basin as precipitation. All the advantages come as
    the water moves across the basin area. One of the advantages of
    the WP is that it allows us to assess the productivity of multiple
    use systems. Multiple uses refer to output such as fish production in
    canals. By-product of agriculture supplied crop residue is an impor-
    tant source of livestock feed and fodder. Basin scale has multiple
    benefits arising from using the same water source. Calculation of
    WP of depleted water in a multiple-use system could be done for the
    entire process of water uses.

Basin level rationale
For increasing WP at the basin level, the rationale lies in the following heads:

a. Limited water resource is the additional water advantage in contrary
    to more available water to users. Increase WP in the upper regions
    of rivers to reduce water depletion and hence increase water avail-
    ability in downstream zone.
b. Reduce water demand and improve water resources, that is, water-
    shed development, groundwater recharge, rainwater harvesting
    through farm pond and tank.
c. Basin development for additional advantages to get productive use
    of the available water resources. For increasing WP to get various
    social, economic, and ecosystem purposes.

7.8 EFFICIENCIES MEASURED IN WATER PRODUCTIVITY

i. **Efficiency**
    It is defined as the output of a specific operation in relation to the
    water use and it is expressed as percentage.

ii. **Transportation efficiency (ηt)**
     It is the efficiency of transportation of water from the source to the
     irrigation dam or draw-off point on the farm boundary.

iii. **Distribution efficiency (ηd)**
      It is the distribution of water from the irrigation and draw-off point
      on the farm boundary through the irrigation system to the point

where it leaves the distributor. Losses from the irrigation dam are included here.

iv. **Conveyance efficiency (ηc)**
It is the combination of the two above and is defined as the efficiency of conveyance of water from the source to the point where it leaves the point of distribution.

v. **Application efficiency (ηa)**
It is the efficiency with which the water leaving the distribution point of the irrigation system falls onto the soil surface.

vi. **System efficiency (ηs)**
It is the efficiency with which water from the irrigation dam or draw-off point on the farm boundary is delivered through the irrigation system to the point where it falls onto the soil surface.

vii. **Storage efficiency (ηo)**
It is the efficiency with which the water that falls onto the soil surface infiltrates the soil and becomes available in the root zone of the plant.

viii. **Field application efficiency (ηf)**
It is the efficiency with which the water leaving the distribution point infiltrates the soil and becomes available in the root zone of the plant.

ix. **Irrigation efficiency (ηi)**
It is the efficiency of the total process of irrigation from the source of the water to the point where the water becomes available in the root zone of the plant.

7.9   WATER PRODUCTIVITY IN MONETARY TERM

PW is the assessment of the economic and livelihood outputs. This kind of outputs could be brick making, crop production, fishing, livestock, and watering. Units are employment per m^3, Rs/m^3, total biomass (kg/m^3), families per command area, and others. The PW is available or received from the intended or unintended products within the total command area. WP is therefore a huge consideration of the products that are received from the irrigation system.

7.10 IRRIGATION PRODUCTIVITY

Irrigation productivity is a calculation of the economic and biophysical advantages. It comes from the use of a unit of irrigation water in agricultural production system. It is expressed in productive crop units of kg/ha, kg/m³, or Rs/m³ etc. This is the product that is obtained from the irrigated crop to which the diverted water was planned. Hence, just to consider product get and capture the whole system of water use and recycling of water. Irrigation included the products from drain water use and rice ratoon products.

The main difference between efficiency and productivity is physical quantities of water, both in the denominator and the numerator. It does not capture differences in the value of water in alternative uses where the gains in basin efficiency can create an important achievement to gains in productivity.

7.11 WATER PRODUCTIVITY MEASUREMENT

This measures in all its uses within a water system. The uses might be included in all irrigation sectors, that is, domestic, industrial, irrigation, fishing, and others.

$$WP = P1 + P2 + P3 + \ldots\ldots\ldots$$

Where
WP = Water productivity
P = Water use productivity
1, 2, 3 = Water utilization

Productivity can be measured in physical, economic, and social aspects. The physical measurements represent as yield of crop per amount of water losses in producing crop yield. Economic measure is the physical output and it is transformed into a value of rupees.

7.12   STRATEGIES FOR ENHANCING WATER PRODUCTIVITY

   i.   The gap between irrigation potential created at the cost of huge amount of national exchequer and utilized as well as gap between generation of water management technologies and its implementation in farmers' field is reduced to increase crop WP and WUE.

  ii.   It is collaboration with private researchers, international and national research institutes, and NGOs for sustainable land and water use and agricultural development. Findings of the study in the pilot areas will assist scientists and researchers to understand the magnitude of arsenic contamination problem. This may further influence agricultural scientists to undertake similar actions in rest of the areas.

 iii.   Scientific dissemination is highly required through participation in conferences, workshops, and seminars. Training and extension materials, that is, courses, brochures, etc will be produced to reach farmers, extension agents, and irrigation policy makers to lead dissemination of the proven water management saving technologies in the state.

 iv.   New academic courses structures should be launched exclusively in collaboration with the Department of Agriculture, government, and the corporate houses including the industries.

  v.   The certificate and diploma courses are offered on WP, water management, water quality management, and crop production. Under this process many users utilize the technologies.

7.13   ACTIVITIES IMPLEMENTATION

   i.   On-farm experimentation to increase WP enhancement, to increase WUE, thereby creating water resources along with quality management for various water to be used for irrigation purposes and their effect must be studied in-depth.

  ii.   Water management research institute or centers like WTC or WALMI to be established more and more; hence, there is no alternative to water management research.

 iii.   Coordination committee to work on the formulation of technical program, generation, and transfer of knowledge to the end users.

iv. The areas like watershed, dry land, rainfed, and wastewater along with work on water quality should be taken on priorities in an integrated approach.

v. The location-specific proven water saving technologies generated over the years should be transferred through single window to the end users with main objective to provide suitable training to the trainers and the farmers on irrigation scheduling to crops for obtaining optimum crop yields with maximum economy of irrigation water;

vi. More need to do for crop field for higher productivity including timely and effective drainage are to be associated to fulfill the other objectives.

vii. The process to formulate the procedures and principles of effective rainwater management for the purpose of irrigation is in the way to adopt.

viii. Determination of moisture dynamics and effective utilization of soil moisture for higher crop production are being calculated for transmitting to the farmers in their languages;

ix. The work to find out the economics and social interpretations of irrigation scheduling to the field crops has been started.

7.14   FUTURE RESEARCH

i.   Evaluation of suitable cropping pattern with diversification of crops in different tube well commands.

ii.   Various irrigation system (drip, sprinkler, and subsurface) on the nutrient dynamics in soil profile.

iii.   The irrigation, crop, water quality, and marketable price to be worked out including chemical analysis of soil, water, and produce and its interrelationships are to be studied.

iv.   Long-term assessment of effect of irrigation on water table, nutrient status, and quality of groundwater.

v.   Economics of the system and cost effectiveness.

vi.   Drainage requirement in different water table areas to study the problem of water logging under lowland ecosystem utilization.

vii.   Training and extension services to the farmers to minimize the gap between technology generation and adoption.

viii. To make popularize the multiple use of water for getting maximum return per drop.

7.15   CONCLUSION

The water input can be specified as volume (m^3) or as the value of water expressed as the highest opportunity cost in alternative uses of the water. There are some other different indicators that could be used to assess water productivity: crop scale, field scale, field farm scale, irrigation system scale, sub-basin, and basin scale. Crop scale is of interest to crop physiologists to assess how efficiently a particular crop is converting water into biomass. Field scale is of interest to the farmer, agronomist, and water specialist to assess the beneficial uses. Field farm scale is used as beneficial and as well as non-beneficial, those are of interest to the farmers, agronomists, and water specialists to assess the opportunities of saving water lost through non-beneficial use. Irrigation system scale is considered by the irrigation system manager in assessing how productively the water available to the irrigation system is being used. Sub-basin scale is of interest to planners and river-basin managers in assessing options for increasing water productivity. The output can be quantified as either biomass or is very difficult to value in monetary terms. River-basin managers and planners in assessing the productivity of the renewable water that enters the basin would depend on mainly rainfall. Calculation of water productivity of depleted water in a multiple-use system could be done for the entire process of water use.

KEYWORDS

- **water productivity**
- **water saving**
- **water use efficiencies**
- **gross inflow**

CHAPTER 8

Surface Water and Groundwater

ABSTRACT

Water resources include information on precipitation, surface, and groundwater. India receives an average precipitation of 1170 mm (46 inches) per year, or about 4000 cubic km (960 cu miles) of rains annually or about 1720 cubic m (61,000 cu ft) of fresh water per person every year. Eighty percent of its area receives rains of 750 mm (30 inches) or more a year. However, this rain is very erratic and uneven that occurs not uniform in time or in a particular location. Most of the rains occur during monsoon seasons (June to September), having about only average of 70 rainy days in a year with the north-east and north receiving far more rains than India's west and south. Other than rains, the melting of snow over the Himalayas after winter season feeds the northern rivers to varying degrees added the water resources. The southern rivers however experience more flow variability over the year. India harnessed 761 cubic km (183 cu miles) (20%) of its water resources in 2010, part of which came from unsustainable use of groundwater. Though the overall water resources are adequate to meet all the requirements of the country, the water supply gaps due to temporal and spatial distribution of water resources are proposed to be bridged up by interlinking rivers project. The total water resources going waste to the sea are nearly 1200 billion cubic meters after sparing moderate environmental and salt export water requirements of all rivers. Food security is possible by achieving water security first which in turn is possible with energy security to supply the electricity for the required water pumping as part of its rivers interlinking from 'surplus' to 'deficit' zones.

8.1 INTRODUCTION

The groundwater goes down through rocks and soil and is stored below the ground. The bedrocks in which groundwater is stored are called aquifers. Aquifers, that is, water-bearing layer are made up of gravel, sand, and sandstone. Water infiltrates and seeps through these rocks due to wide network pore facilitate infiltration. The filling area is called the aquifer or saturated zone. The depth from the surface at which groundwater is found is called the water table. The water-bearing strata may be as shallow deep underground. There the chance of heavy rains can cause the water table to rise. Continuous extraction causes groundwater level to fall down.

The groundwater defines the potential of this resource to irreversible degradation of soil and water.

In India, underground sources of groundwater are in the following categories which are described below:

8.2 BED ROCK OR HARD AQUIFERS OF PENINSULAR ZONE

The water-bearing layer consists around 65% of India. These hard rock aquifers are found in Central India. Land is underlain by hard rock formations below the ground. It was observed that these rocks give rise to a complex and extensive low-storage aquifer system, where in the water level tends to fall rapidly once the water table is down by more than 3–6 m. This aquifer has poor infiltration rate which resists recharge through rainfall due to several reasons. It implies that water in these aquifers is nonreplenishable and may dry out due to continuous usage of groundwater in different sectors.

8.3 ALLUVIAL AQUIFERS

These water-bearing strata, mostly observed in the Gangetic and Indus plains and exist enough storage spaces, and it is the source of fresh water supply. Excessive groundwater extraction and low recharge rates. It is established that the water resource potential or annual water yield of the country in terms of natural runoff in rivers is about 1870 billion cubic meters per annum.

With this scenario of the country, water resources have been estimated as 1130 billion cubic meter per annum. The constraints of uneven distribution of the water resource are difficult in various river basins to the entire available 1870 billion cubic meters per annum. Out of the 1130 billion cubic meters per annum, the part of surface water and groundwater is 690 billion cubic meter per annum and 434 billion cubic meters per annum, respectively. The annual groundwater present for the entire country is 400 billion cubic meter.

The total contribution of rainfall to annual groundwater resource is 68%. The other sectors also share of canal seepage, return flow from irrigation, recharge from open ditches, low-lying areas, tanks, ponds, and water conservation structures contribution together is about 32%.

The groundwater extraction scenario is too high toward a crisis of groundwater. The term groundwater over use or overexploitation is referred to as a situation in which average extraction rate from aquifers is greater than the average recharge pattern (rate). It is a known information that the availability of groundwater is less than that of the surface water. The use of water for irrigation purposes to the decentralized availability of groundwater, it is easily available which is utilizable for and forms the largest share used in the irrigation sector, making it the highest category user in the country. The groundwater is used only 8% for domestic purposes. Industrial use is only 2%. Overall estimation expresses that about 50% of urban water requirements and 85% of rural domestic water requirements are also fulfilled by groundwater source. Major part of irrigation through groundwater is already estimated rather its use increases at increasing rate. Groundwater has big sector to use the water extracted for irrigation. The sources of irrigation are canals, tanks, and wells, including tube wells. Groundwater constitutes the largest share among these sources of water. The human- made source of water like wells, include dug wells, shallow tube wells, and deep tube wells. Over the years, there was a groundwater utilization increasing for irrigation purposes. It was also pointed out the increased usage of groundwater was for agriculture use. The irrigation dependency on groundwater was increased with the onset of the Green Revolution time. For increasing the farm inputs at the same time, water requirement was also more for better utilization of the farm inputs.

8.4 CONJUNCTIVE USE OF WATER

Groundwater caters to more than 45% of the total artificial application of water in India. The groundwater irrigation availability results self-sufficiency in food grains production. Nowadays, the groundwater is an annual rechargeable resource; its availability is nonuniform with time. Judicious use of water in agriculture sector is planning its agriculture sector development. The groundwater development in the eastern part of India estimated is only 18% in Odisha, 21% in Jharkhand, 39% in Bihar, 20% in Chhattisgarh, and 42% in WB. The water availability is more and water inflation is a problem as flood is common phenomenon of this region. The rice is a major crop as per water uptake and also rice is a water loving plant where no option except rice. However, safest water is groundwater for drinking and for irrigation as it is 100% pure as it is sweet too. Except few areas obtaining polluted water, reason is parent material has some heavy metal under the surface.

Annual rechargeable net groundwater resource is about 2,100,928 ha-m of India, out of which 122,126 ha-m is used for domestic and industrial purposes. The present demand for irrigation use is figured out to be 300,901 ha-m. Odisha is a resource-rich state in India. Low crop production and cropping intensity status in Odisha state was also seen. It is important to make use of excess rainwater of wet season judiciously. This excess water can be recharged in to the ground by various recharged techniques, and rainwater harvesting techniques is also one of them. This recharged groundwater can be used during lean period.

Many of the farmer numbers get advantages by growing seasonal crops. During dry season, its depth goes down, while in rainy season, its surface level is very near. The water productivity was more in egg plant crops. The open well is more cheap and profitable and bore well takes few years to repay the cost of installation. The cost-benefit ratio was also very high due to water source and its cost per unit area with two crops as compared with one crop.

8.5 RAINFALL PATTERN AND DISTRIBUTION

The rainfall distribution and pattern of rainfall status in this state of West Bengal; the total rainfall received is 1492.8 mm rainfall. Most of the rainfall, that is, 1295.7 mm (86.8%) is received during June–October, 98.6 mm (6.6%) during November–March, and rest amount of 98.5 mm (6.6%) during summer season.

8.6 EVAPORATIVE DEMANDS OF THE CROPS

The evaporative demand in India is about 1657.8 mm. This crop evaporative demand is more during rain fall received in India. Uneven distributions of rainwater, as well as crop evaporative demand is also uneven. For fulfilling crop water demand during off season, groundwater is needed to improve cropping intensity.

Groundwater balance model is considered for assessing optimal crop planning. The water balance model developed and showed the additional water resources available is 400.84 Mm^3 for subsequent use, is more than present demand due to more groundwater recharge from rainfall and base flow from river or stream. Groundwater is lifted during off season and it is recharged with rainwater during rainy season.

This uses the poor quality water, that is, saline or alkali water for crop production and improved water use efficiency. Due to sodicity in water, the crop suffers with excess salt. Irrigation is needed with permissible limit. This irrigation water pollution (saline and sodic) is available in coastal and northern region of India. These states are Gujarat, Odisha, AP, TN, Goa, Maharashtra, West Bengal, Punjab, Haryana, western UP, and Rajasthan. In eastern India, only saline water is present and good quality water is also available for crop cultivation.

8.7 SURFACE AND GROUNDWATER CONJUNCTIVE USE

Conjunctive use is assumed important as surface and groundwater are correlated. Uneven or aberrant weather condition makes it difficult for the decision maker to derive the water use. Aberrant conditions are efficiently tackled through suitable strategies. Scientists create model for gaining an ideal conjunctive use policy for groundwater withdrawals and spatially distributed cropping system. The under surface water lift were bound to keep the water table elevations within an appropriate range. Scientists developed model to derive the optimal operating policy. A farmer's level program model to run ideal crop plan for an irrigation system for conjunctive use of surface and groundwater has been formulated.

Effective groundwater use in conjunction with available surface water or harvested excess rainwater has a wide use. Water is used in multiple ways under flood-prone environment under low-lying areas of Indo-Gangetic basin to increase crop productivity at the cost of good quality surface water. Multiple uses of water and poor quality water as per water quality can be used in different sectors including fisheries and crop diversification for food security, improved health, and increased employment of resource for poor families especially women folks and better environment in flood-prone ecosystems.

Indiscriminate use of groundwater in agriculture and industrial sectors causes a drastic depletion of water table. In high crop intensification areas, groundwater reservoir replenishes in the first quarter of the year, and later when it goes beyond the lower layer then users wait for annual precipitation for crop production. The recharging rate of water to groundwater reservoir depends on rainfall, runoff, stream flow, and infiltrability of the soils. It also depends on the soil particles, rock, bed rock, gravel, and sand materials present prior to reach the water up to water-bearing strata. Growing of mono-cropping of rice and other intensive cropping system, three to four crops in a year in the same land requires huge amount of water. This water is supplied from groundwater resources. The tube wells, that is, deep and shallow tube wells are dug unjudiciously. The after effects of groundwater deplete rapidly and create severe health and environmental hazards with the contamination of heavy metal in the food chain. The problem could be mitigated by artificial recharging of the groundwater for substantial increase of water table to increase more and more surface moisture for growing various crops under diversifications and suitable sequences and to get as safest sources of arsenic-free water as the possibilities of contamination of arsenic are very less in surface water.

The monsoon starts in the month of June–July and ends in the months of October where rainfall received in north east part and eastern part of India. High rainfall with short time causes floods during rainy season and uneven and aberrant weather conditions. The irrigable of available water resources are unevenly distributed resulting in seasonal deficit of water. Heavy downpour causes devastated flood in some specific area and runoff of the water resources takes place which is of no use. It is of prime importance to store such excess water in the reservoirs for its subsequent uses.

Groundwater recharge can be enhanced by substantially assuring the greater recharge of the water-bearing strata. Maximum water exploitation occurs with deep layer before monsoon groundwater table before the monsoon period. As we know, it should exploit with a limit, but the groundwater extraction should be not more than the annual amount of recharge. Integrated farming system requires more water which needs groundwater exploitation to meet up the irrigation water and agriculture with high-yielding fertilizer and irrigation responsive photoinsensitive short duration rice cultivars that is 3–4 times in a year. Integrated farming does not follow the water use upper limit and posed severe problems of lowering the water table. Heavy metal concentration increases in fresh water and also nitrate leaching takes place. The natural springs have abolished from the farmers' field.

As a result, greater exploitation of groundwater prior to rainy season facilitates water recharge which infiltrate during rainy season. The harvested rainwater is more effective to mitigate the arsenic, mercury, and cadmium like problems which directly or indirectly affects our health. Flood-prone fragile ecosystem is characterized with fertile land and submergence of water from 50 to 400 cm for more than 1 month depending upon the amount and duration of rainfall, and the depth, time, and frequency of flooding. These areas constitute approximately 2.1 M ha area in Eastern IGP, which covers eastern U.P., Bihar, West Bengal, and Bangladesh. System owners are mostly small holders and resource-poor farmers. Mostly, rice is taken with poor productivity once in a year due to uncertainty in timing and pattern of flood. Few mixed crops in summer season, water chestnut, and fox nut in some perennially submerged lands and traditional fish culture in few pockets are taken with low harvest. Poor productivity with underutilized potential of such ecosystem poses a challenge to provide improved livelihood as the resource to poor farm families where mostly male members are migratory laborers and women folks are under employed. Malnutrition and poor health are the consequence of said situation under this fragile environment.

8.8   FARMING SYSTEM AND WATER USE

On station integrated farming system (rice+ mungbean+ sesbania+ fish+ vegetable+ pigeon pea) has been reported in deep low land. The high-yielding variety of rice could not sustain here against local varieties. Other strategies had been focusing mainly on crop-based activities. It needs high light; the problems of said complicated fragile ecosystem. There is an urgent need of evaluation and detailed investigations in this problematic region for further characterization and to take corrective measures. Effective drainage required in rainy season results in hydromorphic condition with multiple use of water may be alternate options.

8.9   INTERNATIONAL SCENARIOS IN WATER USE

The Indian surface water resources, that is, rivers, streams, lakes, and reservoirs are vitally important in every phase of life. Harvested rainwater is used for drinking, irrigation, and thermoelectric power industry. Surface and groundwater is an important part of the water cycle. Groundwater is the part of precipitation that infiltrates. Water in the ground is stored in the spaces between rock particles. Groundwater slowly moves underground, generally at a downward angle.

The rainwater harvesting and ground recharge and its saturation take place below the water table. The ground above the water table may be the dirt and rock in this unsaturated zone contains air. The water-bearing strata of the water table has water that fills the fine pores between rock particles and the cracks of the rocks. Rainwater on the land surface is a vital part of the water cycle which is required in maintaining the ecosystem. Runoff water goes to river and then goes to ocean which is of no use means waste of water if it is stored in rivers, lakes, pond, farm pod, and reservoirs; it facilitates groundwater recharge and agricultural use. The water for use every day comes from these sources of water on the ground surface.

One of the segments of water cycle that is essential to all life on Earth is the freshwater existing on the land surface. Water on Earth surface includes the lakes, reservoirs, ponds, streams, canals. The freshwater refers to water containing less than 1000 mg/L of dissolved solids, most often salt. Ground surface of water bodies are generally thought of as renewable resources, although they are very dependent on other parts of the water cycle. Amount of water in rivers, stream, and lakes is always changing due several reasons, that is, flow of water and exit of water. Water movement into the water bodies may be from rainfall, surface runoff, seepage, and small stream. Movement of water from water bodies like watershed, lakes, and rivers is common phenomenon; in addition to this evaporation, infiltration, utilization is by human being. People get into the act also, as people make great use of surface water.

It was clear for us that during the last ice age when glaciers and snow packs covered much more land surface. Hydrological cycle took place all the time and rain, ice, cloud, and vapor occurs. The topography of the landscape certainly was different before and after the last ice age, which influenced the topographical layout of many surface water bodies as of now today. Glaciers are what made the lakes for huge store house of fires.

The Nile Delta in Egypt exhibit, life can even survive in the desert if there is a limited supply of surface water available. Water on the land surface really does sustain life, and this is as true today as it was millions of years ago. Rainwater harvesting occurs the downward percolation of surface water, even aquifers are happy for water-bearing strata. The fish living in the saline oceans are not affected by freshwater, but without freshwater to replenish the oceans these would eventually evaporate and become too saline for even the fish to survive. The most crystal clear thing is it can be seen that people live near the coasts. The value of lake front property is probably sold for much more than other land.

The streams and lakes are the most important media or phase of the water cycle. Pond, farm pond, ditches also supply the human population, animals, and plants with the freshwater they need to survive. Fresh water dominates only about 3% of all water on Earth and freshwater lakes and watershed account for a mere 0.29% of the Earth's freshwater. Around 20% of all fresh surface water is in one lake which is Lake Baikal in Asia and designated as the largest lake in the world. Remaining 20% near about 5500 cubic miles is stored in the Great Lakes. It can be seen that life on Earth survives on what is essentially only fresh water. In contrary, people on Earth also polluted the fresh water in large quantities in many ways. The systems permit people to live in places where nature does not always supply enough water or where water is not available at the time of year when it is needed. Human being calling the devastation of entire nation, they polluted their fresh water. Nowadays, people using pesticides, fertilizers, industrial effluents, emission of vehicle, nuclear residue, and so many directly and indirectly contaminate our food and drinking water resulting in many diseases

8.10 MODE OF DISTRIBUTION OF EARTH'S WATER

There is plenty of water in the Earth, about 71% of the Earth's surface is water covered and water in different phase, which is conserved as water in under surface, that is, water-bearing layer or strata.

   a. Ocean water is saline and occurs about 97.5% of Earth's total water. Very small amount of water that is actually sweet water. Fresh water is used for human animal and agriculture use.

   b. Earth's water contains only 2.5% which is fresh water is used for human beings and wild animals.

   c. Most of the freshwater is locked up as ice and in the underground.

   d. The surface freshwater about 20.9% of 2.5% is found in lakes. Rivers make up 0.49% of surface freshwater and rivers account for only a small amount of freshwater.

Earth has plenty of water for human, plant, and animals. The water underlies the Earth's surface almost everywhere, beneath hills, mountains, plains, and deserts. Accessible fresh water is enough for use without treatment, and it iss sometimes difficult to locate or to measure and describe. The land surface also has marsh, or it may lay many hundreds of feet below the surface.

8.11   CONCLUSION

Successful agriculture is dependent upon farmers having sufficient access to water. However, water scarcity is already a critical constraint to farming in many parts of the world. World Bank targets food production and water management as an increasingly global issue that is fostering a growing debate. Water scarcity is where there is not enough water to meet all demands, including that needed for effective functioning of ecosystems. Water scarcity also occurs where water seems abundant but the resources are over-committed. Physical water scarcity also includes environmental degradation declining groundwater resources. Starvation and poverty has got nexus with water availability and economic scarcity, meanwhile, is caused by a lack of investment in water or insufficient human capacity to satisfy the demand for water. Symptoms of economic water scarcity include a lack of infrastructure, with people often having to fetch water from rivers for domestic and agricultural uses.

KEYWORDS

- **surface water**
- **groundwater**
- **conjunctive use**
- **aquifers**
- **rainfall and ET demand**

CHAPTER 9

Measurement of Irrigation Water

ABSTRACT

The net quantity of water to be applied depends upon the magnitude of moisture deficit in the soil, leaching requirement, and expectancy of rainfall. When no rainfall is likely to be received and soil is not saline, net quantity of water to be applied is equal to the moisture deficit in the soil, that is, the quantity required to fill the root zone to field capacity. The moisture deficit (d) in the effective root zone is found out by determining the field capacity moisture content and bulk density of each layer. Soil moisture is estimated both by direct and indirect methods. Direct methods involve the determination of moisture in the soil while indirect methods estimate the amount of water through the properties of water in the soil. In indirect methods, moisture is estimated by the thermogravimetrically either through oven drying process or by volumetric method.

9.1 INTRODUCTION

Water is a precious commodity; its proper utilization is very much essential to maintain water balance in the environment and equal distribution to all creatures. Maximum water is utilized in irrigation purposes. Water's application efficiency is very much needed. Total water requirement has to be estimated to provide other crops in other season during the year. Irrigation water is measured in different farm field and farmer's field to estimate the water requirement crop wise. As per water availability and its demand in different phase of crop, it is decided to apply in proper manner to harvest the crop without water stress. So, measurement of irrigation water is important in all aspects.

9.2   REQUIREMENT OF MEASUREMENT OF IRRIGATION WATER

1. Estimation of the amount of water supply to the crops.
2. Estimation of runoff and steam flow is essential workout in planning suitable soil and water conservation measure.
3. Estimation of well performance for crop yield.
4. Estimation of irrigation water is needed for field experiments.

9.3   UNITS OF MEASUREMENT

Flowing water in a channel or pipeline are m³/s, lps, ft³/s, and gpm. Water at tank, reservoir, or standing water in field are measured by cubic meter, liter, and centimeter, cubic feet, gallons, acre-inch, acre-feet, etc.

9.4   MEASUREMENT OF IRRIGATION WATER

Many devices are generally used for measuring irrigation water in agricultural field. Those devices most commonly used are weirs, orifices, Parshall flumes, calibrated gates, and rating flumes. In addition to the foregoing devices, current meters, Clausen-Pierce Weir gauges, and slope area methods are used by expert who is technically trained. Many mechanical devices for measuring flow have been developed as per need, some of which not only measure rate of flow but also register the total volume of water passing during any period of time.

9.4.1   CURRENT METERS

The water discharge of a river can be estimated directly by means of the velocity and the cross-sectional area of the channel. The most accurate method of determining the velocity of the water is by the use of a current meter, of which several types are available. Current meters have a wheel which is fitted with cupped vanes. It is mounted with an axis; it is free to move under water pressure. This wheel is on the up end of a horizontal shaft that has its other end fitted with directional vanes. It is used in shallow water where it is possible by a vertical rod held in the hands of the observer. In deeper water situation, the rod is removed and a cable is attached to the meter. To the lower side of the meter is attached a streamlined lead weight which steadies the meter and holds it in position. The wheel revolves at a rate proportional to the velocity of the current in which it is placed. The immersed revolutions of the wheel are counted by an electric current. It is detached at each turn by means of a commutator. The electric current passes through the meter, tiny buzzer records the break of the circuit. Current is generated by a small dry cell battery installed with it. The stream out flow is the product of its cross-sectional channel. The average velocity is measured of the water passing through the given cross-sectional channel. Water meter consists of accuracy value of two factors, that is, velocity and cross-sectional area or height of it. The distance of each point is measured from a permanent point on the bank. Best results are obtained from gauging, that is, vertical strips, preferably of equal width and discharge.

9.4.2 FLOAT MEASUREMENTS

A rough estimation of the water flowing in a straight uniform channel, some object such as a piece of wood or an apple may be thrown into the stream and the time required by the object to go a known distance downstream is noted. The height of the water is measured with length in square feet multiplied by the surface velocity in feet per second. It is estimated by timing of float, multiplied by a coefficient to correct for the fact that the surface velocity of the water is greater than the average velocity, it will give on an average discharge for the stream. Also it differs widely, depending upon the shape of the cross-section and the condition of the banks and bottom of the channel; the average coefficient is about 0.85. The discharge therefore equals approximately 85% of the cross-sectional area. The float method is not advisable where other more reliable means are not available. The float method is the simplest, cheapest, and reliable devices weir.

i.   **Weir A:** Weir A is part of a device which is bulkhead placed across a ditch or stream with an opening cut in the top through which the water is allowed to pass. The opening of this device is called the weir notch.
ii. **Weir Pona**: Weir Pona is part of the device and it is fixed at ditch immediately upstream from the weir.
iii. **Weir T Crest:** Weir T crests is also a part of the device which is fitted at the bottom of the weir notch.
iv. **Head on Crest:** The height of water flowing is estimated at some point in the weir pond.
v.   **Sharp-Crested Weir:** A weir having a thin-edged crest and sides such that the overflowing water touches the crest at only one point.

vi. **End Contraction:** The horizontal distance from the end of the weir crest to the side of weir panel.

vii. **Bottom Contraction:** Bottom contraction is part of the device which is set at a vertical distance from the weir crest to the bottom of the weir pond.

viii. **Weir Scale Gauge:** The attached gauge and its scale are fastened on the side of weir to measure head and crest. This weirs scale gauge is divided into two general classes; that is, sharp crested and broad crested. The crested is classified into two, that is, end weirs contractions and weirs without end contractions. These weirs are made of steel sheet or wood and are placed in the ditch where a measurement is taken. Among these, stationary structures may be built of wood, steel sheet, and concrete. However, in wood and concrete design, the notch is usually faced with a metal strip which is made up of the sharp crest. A weir notch gives discharge in a proportion to the head on the crest and is affected by the state of the crest, the connection, and the velocity.

9.4.3 ANGLE IRON CRESTS RECTANGULAR WEIR

This device properly measures water with a rectangular weir, constructed and installed in which the formula, tables, and curves were developed. Some general requirements for the proper setting and operation of weirs are given below:

i. The weir sets at the lower end of a river long, wide, and deep, smooth current with a velocity of 0.5 ft/s.

ii. The longitudinal axis of the weir should be perpendicular to the direction of the flow of water. The centerline of the weir is attached parallel.

iii. The weir is vertical, and right angles to the direction of flow of water.

iv. The weir is horizontal and the water passing over it will be the same depth at all points.

v. The distance covered by the crest above the bottom of the river or stream is three times the depth of water. The sides of the river are at a distance covered from the sides of crest.

vi. The rectangular gauge is placed on a stake at any point in the weir pond. The weir box supply sufficiently above or to one side of the weir. The weir scale is far enough to one side. The level zero of the weir scale or gauge must be placed at with an actual position. This may be prepared with an ordinary carpenter's efforts or, where greater refinement is desired at engineer's level.

vii. The estimation of the crest head or depth of water on the crest may also be made by placing scale on a lug to the side of the weir notch or on a stake placed in the weir pond 4 or 5 ft above the rectangular weir. The stake must be placed at a level with the rectangular weir crest.

viii. The rectangular crest moves freely below the weir. With this subsequent, it leaves an air space under the over falling sheet of water. The water level must be below the weir rises above the crest elevation. These are made during measurement on submerged weirs and are unreliable under this situation.

ix. It depends upon the depth of water and it lies more than 2 cm. The water tends to cling to the downstream side of the rectangular crest and the relationship between the depth of water on the crest and the discharge no longer holds for measuring water.

x. Drawback regarding erosion also noticed to prevent erosion below the rectangular weir; the ditch downstream should be protected by loose rock or by mud.

9.4.4   RECTANGULAR WEIRS AND ITS LIMITATION

Rectangular weirs are easy to construct and convenient to use. The disadvantages of weirs are that they require a considerable loss of head which may not be available on ditches with a flat grade attachment. The weir pond causes deposition of silt from those streams carrying a heavy silt load or mud and these deposits in the channel of approach destroy the proper conditions for weir measurements. These weirs should not be combined with headgate structures of the device for accurate measuring.

9.4.5   DIFFERENT WEIRS STRUCTURE

i.   Veil Structure

These weir structures may be either portable or stationary with standard for easiness. The metal weirs are more satisfactory than wooden weirs for portable use. It is made of galvanized iron stiffened by means of heavy angles.

Veil structure with a simple weir bulkhead and weir pond may be used or a weir box may be made. In clay soils, is needed washing to keep smooth run, a simple bulkhead can be used to an advantage, but in light soils, which is subject to easy erosion, a weir box with wing walls and cutoff walls should be used to prevent the structure from being washed out. The head is measured from a stake in the weir pond set at the same elevation as the crest.

Use of weir

a. Weir immediately below a curve in the ditch at the curve will cause the water to flow.
b. Do not set it immediately below or too close to a headgate where the water has high velocity.
c. Do not permit the water below the weir as it will not allow complete contraction and it will discontinue the discharge.
d. Do not set the weir any other way than vertical and at right angles to the flow of the stream.
e. Do not attempt to use too small a weir. Keep in a larger weir where the water to be measured exceeds a depth on the crest of one-third the crest length.
f. Do not allow the pool above the weir to fill up with sediment and increase the velocity of water.

9.4.6   MAKING A DISCHARGE MEASUREMENT WITH A WEIR

Measuring the depth of water over the crest of the weir is made by placing a carpenter's rule on the stake which has been fixed at the elevation of the crest or by reading the weir scale. The curves are so constructed that the head may be measured either in feet or in inches and with the crest lengths varying from 6″ to 4 ft. For calculation, suppose that the head on the crest is 2 ft, rectangular weir measures 8.5″.

i.   Rectangular Weir

The rectangular weir is named after the shape of its notch shape. The rectangular weir is the oldest form of weir used worldwide. The rectangular weir is simple, easy to construct, and accurate, when properly installed and used. The recommended size of rectangular weirs varies in size and streams of water. The maximum and minimum discharge for the various sizes of weirs overlap. The results from the curves shows that the discharge for weir crests of 1.0, 1.5, 2.0, 3.0, and 4.0 ft for heads height varies from 0.1 ft to 1.5 ft.

ii. Trapezoidal or Cipolletti Weir

Italian engineer designed the trapezoidal or Cipolletti weir. It is used extensively in irrigation work. It gives equally accurate measurements but is more difficult to construct than the rectangular weir. The trapezoidal weirs and its side slopes are uniform and have an inclination of one horizontal to four vertical slopes. The discharge of water may be considered in two parts, one through the rectangular area of the length which is equal to the crest length. The second one passes through a triangular area equal to the sum of the area of the two triangles formed at the ends of the rectangle state. The total discharge of water is therefore greater than from a rectangular contracted weir having an equal crest length. The length of crest; that is, a trapezoidal weir having a crest length of 4 ft discharges twice as much as one having a crest length of 2 ft. The generally recommended size of trapezoidal weir is to use with various size streams of water. The point to be noted is that the property of equal discharge per unit length of weir crest makes the trapezoidal weir an excellent device for use in division of amount of water in addition to being a measuring device. The water discharge curves are obtained for trapezoidal weirs having crest lengths from 6" to 4 ft.

iii. Ninety Degree Triangular Notch Weir

This weir is especially suited to small quantities of water of a second foot to about 2.5 s ft. The 90° makes an angle of 45° or half pitch with the vertical. Discharge limit is maximum of 2.5 s ft.

iv. Rectangular Suppressed Weir

The rectangular suppressed weir or weir without end contractions is made of a bulkhead in a rectangular flume section. The bulkhead should be sufficiently high from the bottom of the flume so that the distances from the weir 36 crest to the bottom of the flume is at least twice and mostly three times. The flume section should be uniform in cross section, having a horizontal base. The bulkhead should be set in a vertical plane, and the upstream face is smooth with a crest width not in excess of W' in thickness. The head is determined in the same manner as for other weirs. For more accurate outcomes, a stilling well and a hook gauge are recommended. The suppressed weir completely fills the flume, and cuts off the free circulation of air under the over falling sheet. It is easy to function, artificial ventilation should be provided by drilling a small hole in each side wall near the downstream edge and a little below the weir crest and discharge curves for the rectangular suppressed weir are obtained. For example, if the depth of water flowing over the crest of 1 ft with high weir is about 0.75 ft, discharge curves are obtained.

v. **Rating Flumes**

There are many streams having the steep slope, gravel and debris make use of weirs or orifices impracticable. The flow of such streams may be measured by constructing a rating flume and calibrating it by determining the relationship between the discharge and depth of water in the flume. The rating can be done by measuring the water at various gauge heights with a current meter. The services of a competent engineer should be sought to rate the flume. The device is calibrated and the discharge is estimated reading the gauge. Some problems reported regarding deposits of silt, growth of weeds, or other. The Parshall flume is a unique permanent rating flume used widely.

vi. **Submerged Orifices**

For sections it is difficult to get the required head for flow over a weir and where the waters carry considerable silt, a submerged orifice is sometimes used. There is orifice which is opening cut in a bulkhead through which water flows. The opening is below the water surface on both sides of the bulkhead. The water surface on the downstream side is below the opening. It is said to have a free discharge. The downstream water surface is placed on the top and bottom of the orifice. This condition should be avoided to consider as it is more suitable for common use under the heads available.

Submerged orifices are classified into two:

1. Orifices having fixed dimensions and
2. Orifices having height of opening is varied.

A standard submerged orifice has fixed dimensions with all sides. The opening rectangular and adjustable submerged orifice is one in which the height of opening and head may be varied to fit the conditions. It is usually built with suppressed side contractions of the devices. The ordinary form is the simple headgate. The standard submerged orifice is the more reliable, easy, and handy of these two types.

iii. **Submerged Orifice with Fixed Dimensions**

The amount of water that passes through a submerged orifice of fixed dimensions is estimated by the difference in elevation of water surface upstream and downstream from depth. There is one scale which should be placed on the upstream side of the orifice and one on the downstream side with the zero end of each scale at the same level near the top of the structure. The head on the orifice is equal to the difference in the scale readings which is taken for easy, handy, and reliable. Hence, the size opening should be adapted to the head available and the size of stream. Some rules of this orifice for installing submerged orifices are given below:

a. The orifice is sharp and smooth in upstream. The distance of each from the bounding surfaces of the channel of the upstream and on the downstream side is not less than twice of the orifice.
b. The upstream face of the orifice is vertical.
c. Both side edges are maintained levels from end to end.
d. The head on the orifice is the actual difference and the water surface on the downstream side.
e. The cross-sectional area of the water prism for 20 to 30 ft from the orifice, on the upstream and on the downstream is six times of the cross-sectional area.
f. Correction should be made for velocity of approach where appreciable errors are caused by neglecting the head due to it.

Advantages and disadvantages

The main advantage is that it can be used on relative level canals where it is not possible to obtain sufficient fall for weir measurements. The disadvantages of this orifice are the weir-collecting floating debris, sand, and sediment and the pond in front of the orifice is allowed to silt up.

9.5   DETERMINATION OF DISCHARGE

Discharge curves can be used directly only in connection with an orifice having one of the eight cross-sectional areas given 0.25, 0.5, 0.75 of a sq ft, and so on up to 2 sq ft. It is easier and comfortable, if an orifice is designed and its cross-sectional area is equal. Orifices and their discharge having cross-sectional areas other than those given by the curve should be computed by proportion. For example, the discharge equals 3.78 ft³/s for an orifice area of 1.0 sq ft and a head of 0.6 ft; when the orifice area was 0.6 of a sq ft, the discharge would be $0.6 \times 3.78 = 2.27$ cu ft/s.

9.6   COMBINATION HEADGATE AND MEASURING DEVICES

A constant head, adjustable orifice turnout; it replaces the common turnout gate weir combination. This device was designed to control and accurately measure irrigation water without the excessive amount of adjustment and walking usually required for the gate weir combination. The 18 × 24"and 24 × 30" gates are used as orifices. These gates are used in single barrel and double barrel turnouts. The orifice is designed to operate at 0.2 ft effective head, which is adjusted by the turnout gate after the orifice gate is opened to a height determined from tables compiled for various discharges and gate openings.

i.   Commercial Gates

There is another combination of headgate and measuring device which was proved successfully and it is known as a commercial gate. The principle of operation of these gates is similar to that of the submerged orifice or to the double orifice gate as described. This commercial gate is widely used and adapted for use at lateral outlets for they serve as headgates and it is also used for water measuring instrument. The installation of the gate is perfect for a lateral outlet through a canal bank. In installing, care must be taken to see that the outlet pipe is low enough to insure the proper submergence of the outlet, which should not be less than 6" under lowest conditions. Two measurements are necessary for finding the discharge: the amount of gate opening as found by measuring the length of rod coming through the handwheel and the difference in the elevations of the upstream and downstream water surfaces. Knowing these two measurements, one may enter tables or curves supplied by the manufacturer of the gate and find the discharge. Discharges from approximately 1 quarter second foot to 78 s ft may be measured.

ii.   Parshall Measuring Flume

The Parshall measuring flume has a converging inlet section; a throat section with straight parallel sides. The Parshall measuring flume is a water measuring device by which the water flowing in an open channel can be measured satisfactorily with a minimum loss of head. The loss of head for the free flow limit is only about 25% of that for the over pour weir. The accuracy of discharge measures with this flume under normal operating conditions. The Parshall flume operates as a free flow single-head device or under submerged flow conditions where two heads are involved. There are only the free flow single head devices which are considered. The dimensions of flumes are 0.35 ft³/s to 176 ft³/s. The Parshall flume consists of a box of wood, concrete, or metal with a level floor, a converging section, and vertical side walls. At the end of the converging section, the floor slopes downward 9″ to 2 ft. The outlet section is 3 ft long and diverges. The floor of the outlet section rises 6″ in 3 ft. This device has a dimension, that is, length and width for attaching a throat width of 1 ft or more. There is a need of flow conditions in this device; a throat has width of 1 ft or more; for smaller flumes, such as a 3″ flume, the submergence should not exceed 50%.

The submergence in percent means the percentage that the downstream head is of the upstream head. This requires that the proper size flume for the conditions present be selected. To find the smallest size flume necessary to measure a discharge of 5 s-ft with 62% submergence and with a loss of head not exceeding one-half foot. Enter at the lower left and follow vertically on the line 62 until the curved discharge line 5 is reached. At this point, move horizontally to the right until the vertical line 0.50 is intersected (Skogerboe, 1967).

Note that this point is just a little to the right of the diagonal line marked 1 ft throat width. This indicates that a flume has width of just a little less than 1 ft and has throat widths from 6″ to 10 ft for discharge.

9.7 QUANTITY OF IRRIGATION WATER

The quantity of water depends on moisture deficit in the soil. It is also depends on leaching requirement and rainfall. Under dry situation or no rain received area shows the more saline, net quantity of water to be applied is equal to the moisture deficit in the soil. The moisture deficit (d) in root zone is found out by estimating the field capacity and bulk density.

Problem: Find out the net quantity of irrigation water to wheat field with the moisture regimes.

Sr. No.	Depth of soil layer (cm)	Moisture % on oven dry basis		Apparent-specific gravity g/cc
		Field capacity	Actual	
1	0–15	25.0	17.8	1.39
2	15–30	24.0	17.8	1.47
3	30–60	22.3	19.2	1.51
4	60–90	22.2	20.5	1.53

Solution: Layer-wise moisture deficit will be as follows:

1. First Layer = $\dfrac{25.0 - 17.81}{100} \times 1.39 \times 15 = 1.79$

2. Second Layer = $\dfrac{24.0 - 17.8}{100} \times 1.47 \times 15 = 1.36$

3. Third Layer = $\dfrac{22.3 - 19.2}{100} \times 1.81 \times 30 = 1.40$

4. Forth Layer = $\dfrac{22.2 - 20.5}{100} \times 1.53 \times 30 = 0.78$

The net quantity of water is required 5.33 cm.

9.8 IRRIGATION WATER REQUIREMENT

a. Water measuring units

Water is measured in two ways; these are rest position and running positions. Water is measured in rest conditions such as volume, liter, cubic meter, hectare meter. Water in running positions is measured as liters per hour and meters per day.

i. Liter

Liquid measuring unit is liter and 1 L is equal to 0.22 imperial gallons or 0.0353 ft^3 or 1/1000 m^3.

ii. Cubic meters

For measuring volume of liquid, cubic meter is generally used and volume of water equivalent to of 1 m^3 in length, 1 m in breadth, and 1 m in thickness. If it is expressed in detail, it will be as 1 m^3 of water equal to 1 kL or 100 L or 220 gallons or 35.3 ft^3 or 1 ton.

iii. Gallon

For large volume of liquid, measurement gallon is generally used and 1 gallon is 0.1602 ft^3. It is very common in many countries; 1 gallon of water weighs about 10 Ib.

iv. Cubic foot

Sometimes liquid is also measured as cubic foot and a volume of liquid is equal to that of a cube 1 ft in length, 1 ft in breadth, and 1 ft³ in thickness. One cubic foot of water is equal to 28.37 L or 6.23 gallons or 0.0283 m³ or 0.028 ton.

v. **Hectare centimeters**

Hectare centimeter is usually used to measure the water in large area in agricultural field. Hectare centimeter of water means the volume of water that is needed to spread an area of 1 ha surface to a depth of 1 cm. For easy understanding, 1 ha centimeter is equivalent to 100 m³ equal to 100,000 L of water.

vi. **Acre inch**

Acre inch also measures water in large area such as agriculture field. The water requires covering 1 acre or 43,560 sq. ft area surface to a depth of 1″. One hectare inch is equivalent to 3630 ft³ or 101 ton.

vii. **Acre foot**

Acre foot also measures in large area of agricultural land. The water required to cover 1 acre area to a depth of 1 ft. Here, 1 acre feet equals to 43,560 ft³.

viii. **Cusec**

Cusec is the quantity of water flowing at a rate of 1 ft³/s and hence, the weight of 1 ft³ is about 28.37 kg.

One cusec of water is equal to 62.4 lb × 60 × 60 = 22,464 gallons or 101 tons, or 1 ha inch or 28.37 L × 60 × 60 = 101,952 L or 1 acre inch.

ix. **Duty of Water**

The number of acres covered by one cusec of water flowing continuously to the crop through the season.

x. **Delta**

Delta denotes that the water exists around rhizosphere. It is the total depth of water uptake by a crop.

Problem 1: An average discharge of a pump is 15 L/s to irrigate 1 ha mustard crop in 12 h. Calculate the average depth of irrigation given to the crop?

Solution:

$$\text{Discharge in 12 h} = 15 \times 60 \times 60 \times 12$$
$$= 648,000 \text{ L}$$
$$= 648 \text{ M}^3$$

$$\text{Depth of irrigation (cm)} = \frac{\text{Volume of water (Cu. m)}}{\text{Area of land (sq. m)}} \times 100$$

$$= \frac{648}{10,000} \times 100$$

$$= 6.48 \text{ cm}$$

Problem 2: Maize crop requires 40 cm of irrigation water during 120 days crop period. Calculate the area that can be irrigated with a flow of 20 L per s for 12 h a day?

Solution:

$$\text{Total discharge during irrigation period} = \frac{20 \times 60 \times 60 \times 12 \times 120}{1000} \text{ M}^3$$

$$= 103,680 \text{ M}^3$$

$$\text{Irrigation requirement per hectare} = \frac{40}{100} \times 10,000 \text{ M}^3$$

$$= 103,680 \text{ M}^3$$

$$\text{Irrigation requirement per hectare} = \frac{40}{100} \times 10,000 \text{ M}^3$$

$$= 4000 \text{ M}^3$$

$$\text{Area irrigated} = \frac{\text{Volume of available water}}{\text{Volume of water required/ha (M}^3)} \times 10,000 \text{ M}^3$$

$$= \frac{103,680}{40}$$

$$= 25.92 \text{ ha land can be irrigated}$$

9.9   IRRIGATION WATER MEASURING DEVICES

For measuring the irrigation water there are many devices which are commonly used. They are grouped into four categories:

1. Volumetric measures.
2. Velocity area methods are two types, these are float method and water meter.

3. Measuring structures are three types, that is, a) orifices, b) weirs, and c) flumes.
4. Tracer methods.

1. Volumetric (container used)

A small irrigation river is used for measuring volume and to collect the flow in container of known volume for a measured period. For measures, bucket or barrel is used as container and to fill the container, time is recorded with a stopwatch or with seconds on wristwatch. Thereafter, the rate of flow is measured as under:

$$Q = \text{Discharge rate liter/second} = \frac{\text{Volume of container (liters)}}{\text{Time required to fill (seconds)}}$$

Problem: If 20 L bucket is filled in 10 s by Persian wheel then what will be the rate of flow?

Solution:

$$\text{Discharge ratio liter/second} = \frac{20}{10}$$

$$= 2.0 \text{ L/s or } 144 \text{ L/min}$$

2. Velocity

a. Float method of irrigation water measurement

It is used for rough estimation of the flow in a channel. It consists of nothing; the rate of flow of a floating body. A long-necked bottle partly filled with water or black wood, and then an orange or lemon may be used as float. One channel selected measures about 30 m long with uniform cross-section is selected. Here an attachment of a string is stretched across each end of section at right to the direction of flow of water and there is a device float placed in the channel, a short distance upstream. The float is required to pass from upper end to lower end of the section and flow is estimated. Many trials are made to get the average time of travel.

For estimation of the velocity of water, the length of the trial section is divided by the average time taken. Hence, the velocity of the float on the surface of the water will be greater than the average velocity of the stream. It is a constant factor about 0.85. To get the rate of flow, this average velocity is multiplied with sectional area.

Discharge = area × velocity

$$Q = A \times V$$

Here,
Q = rate of discharge is m3/s.
v = velocity is m/s
a = cross-sectional area is m².

b) Water meters for irrigation measurement

Water meters are multiblade propeller made of metal, plastic, or rubber. This blade rotates in a vertical or horizontal plane. The measurement of water with desired volumetric units, water meters are available. There are basic of the water meter given below:

1. The pipe must flow full at all times.
2. The flow rate is more than the minimum range.

Meters are calibrated and field adjustments are done in factory. The water meters are installed in open channels where meters are set through the pipes of known cross-sectional area. Weeds, grasses, and debris or other foreign materials obstruct the propeller.

9.10   WATER MEASURING INSTRUMENT

a.   Orifices

Orifices are installed in an open channel. It is circular or rectangular openings in vertical bulk head through which water flows. The opening part is sharp and is made of metal. The cross-sectional area of orifice is small to the cross-section of the device. It operates under free flow or submerged flow conditions. The types of orifices are:

I)   **Inlet of orifice**

The discharge of orifice situated below the level of inlet is calculated by the following equation.

$$Q = Ca \times \text{under root } (2gh)$$

Here, Q = Quality of flow in C. ft/s.
     a+ Cross-sectional area in the canal or orifice in sq. ft.
     c = A constant which ranges from 0.6 to 0.8
     g = Accelerated gravity (32 ft/s/s)
     h = Height of water level.

II) Discharge at higher level than inflow pipe

The position of orifice is situated at a higher level than the inflow pipe; the discharge is estimated by the following equation:

$$Q = CLh1.5$$
Here, D = Length of orifice in feet and
$$C = 2C\ 2g$$
H = Head of water.

III) Discharge by submerged orifice

The discharge by a submerged orifice is given by the following equation:

$$Q = 0.61\ LH\ 2gh$$
For example, if L = 1.0 H = 0.5ft. h = 0.25 ft. and g = acceleration due to gravity (32ft/s/s) then Q = 1.22 C, ft/s.

b. Discharge by Weirs

A notch is attached with weirs in a well-built across a stream. Notch of the weir are of different types. These are given below:

a. Rectangular
b. Trapezoidal and
c. 90°V notch

a. Rectangular weir

The length of a weir is equal to width of the upstream channel or less than it. The discharge by rectangular weir, in case of complete end, contraction is given by following equation:

$$Q = 3.33\ (L\text{-}0.2\ H)\ H^{1.5}$$
Here, L = measured length
L = effective length (feet)
L = L-0.2H)

b. Trapezoidal weir

The discharge of water is given by following equation:

$$Q = 3.367\ LH\ ^\wedge\ 1.5$$

$$\text{Where} = \frac{L1 + L2}{2}$$

c. **Discharge through 90°V notch**

The discharge is given by following equation:

$Q = 2.49\ H \wedge 2.48 = 2.5\ H^2.5$

9.11   PARSHALL FLUME

Parshall flume is a device for measuring the loss in the head caused by forcing a stream of water. The accuracy in the Parshall flume is allowable limits of 5%; the flumes ranging from 3″ to 10 ft throat width are used. The flumes of 3, 6, and 9″ size are usually used. This ratio should not exceed 0/3 for 3″, 6″ 9 sized Parshall flumes.

9.12   CUT THROAT FLUMES

Cutthroat flumes are used for the measurement of water and there is no throat section. The flumes have a level floor as opposed to the inclined floor in the throat and exit section in the partial flumes. Flume has the same all lengths in both the entrance and exit sections and these flumes consist of converging inlet section and diverging outlet section. The discharge Q by a cutthroat flume depends upon the upstream depth of flow Ha. The basic form of the free-flow equation is given below:

$$Q = CH\ a\ 1.56$$
$$\text{Here, } C = 3.50\ W\ 1.025$$

9.13   TRACER METHODS

This method is independent of stream cross-section and is suitable for field measurements without installing fixed structures. Tracer methods, a substance in concentration form is flowing water. The concentration of the tracer is calculated at downstream part. The quantity of water is needed to complete the dilution involved. There is no need to measure velocity, depth, and head, and cross sectional. Tracer is measured by the following equation:

Qt = ad

Where Q = Size of stream (liter/second)

t = Time of application (seconds or hour)

a = Area (sq. m or hectare)

d = Depth in cm of irrigated land irrigated

9.14 CONCLUSION

Irrigation water measurement methods should be relatively simple that are are required to be designed which are very simple in operation, inexpensive, most practical and provides reasonably accurate results under most of the situations, those could be used as famers' approach and be widely used by the farmers. Most of the agricultural operations are under limitation concerning the range of moisture measurement (0 to <100 bars). However, it is physically difficult, relatively slow sampling (time-consuming) at higher depths they became labor-intensive, destructive, also having repeated measurement at one spot each time is not possible, faced difficult in gravel and rocky soils. The electric drying oven, soil sampling equipment, and precision balance are required for implementation of such important parameter which are not possible to be afforded by the farmers.

KEYWORDS

- **water measurement**
- **current meter**
- **float method**
- **tracer**
- **Parshall flume**
- **weirs**

CHAPTER 10

Water Quality in Agriculture

ABSTRACT

The parameters of water quality, namely, pH, electrical conductivity, alkalinity, nitrate, phosphate, sodium, calcium, magnesium, TDS, and TSS, have effect or adverse effect on irrigated agriculture. Impact assessment of adequate supply of usable quality water for the purpose of irrigation to influence the crop quality and marketability has a great deal with present art of agriculture. A number of water management technologies for various crops have been developed over years to enhance agricultural productivity as well as profit. The area of crops under irrigation was also increased substantially. The irrigation potential was created at the cost of huge amount of money. There is a large gap between the irrigation potential created and utilized in agriculture sector. The water management technologies developed so far were not also being practiced at the actual users end. There are some unidentified gaps in the existing system, which act as constraints on adoption of water management technologies. Appropriate interventions required to provide corrected measures to address the problems at the farmers field on adoption of water efficient management practices for growing field crops. The influence of irrigation water quality on crop quality, marketability, and mitigation options of adverse effect of unsuitable contaminated polluted waste water on agricultural produces needs more attention at present day of precision agriculture to maximize the net returns on sustainable basis.

10.1   INTRODUCTION

The parameters of water quality, namely, pH, electrical conductivity, alkalinity, nitrate, phosphate, sodium, calcium, magnesium, TDS, and TSS, on irrigated agriculture. Impact assessment of adequate supply of usable quality water for the purpose of irrigation to influence the crop quality and marketability has a great deal with present art of agriculture. The irrigation potential was created at the cost of huge amount of national exchequer and are not used judiciously and a large gap existed between the irrigation potential created and utilized. The influence of irrigation water quality on crop quality, marketability, and mitigation options of adverse effect of unsuitable/contaminated/polluted/waste water on agricultural produces needs more attention at present day of precision agriculture to maximize the net returns. Adoption of location-specific appropriate water-saving technologies including adopting of microirrigation systems would create a favorable environment under sustainable agriculture.

10.2   DEFINITION, CONCEPT, AND IMPORTANCE

It refers to the chemical, physical, and biological properties of water. Water quality in agriculture is an assessment of the state of water relative to the need of one or more biotic species. There are most common standards used to define water quality with environment of ecosystems, human safety, and drinking water.

10.3   IMPORTANCE OF THE WATER QUALITY PROBLEM

The different water quality parameters are pH, electrical conductivity, alkalinity, nitrate, phosphate, sodium, calcium, and magnesium, total dissolved salt, and total soluble solids. The water quality depends on different water quality parameters. Impact assessment of adequate supply of usable quality water for the purpose of irrigation to influence the crop quality and marketability has a great deal with present art of agriculture. The influence of irrigation water quality on crop quality, marketability, and agricultural produces needs more attention at present day of precision agriculture to maximize the net returns.

10.4 IMPACT OF WATER QUALITY ON HUMAN HEALTH

Contaminants present in untreated water include microorganisms like viruses, protozoa, and bacteria. Some inorganic contaminants, that is, salt and metals are present in untreated water and organic contaminants from industrial effluents and pesticides. The very known idea of water quality depends upon the local geology and ecosystem. The strict regulations may be establishing limits for contaminants in packaged water. It was also observed that bottled water which is used for drinking purposes may be expected to retain small amounts of contaminants which may affect food chain as well as human health. All contaminants may not indicate that the water poses a health risk of any age except heavy metals and maximum acceptable limit (MAL) of contaminants in drinking as well as irrigation water.

In urban water, purification occurs in municipal water systems to remove contaminants from the surface water, that is, river, lake, or reservoirs or groundwater before it is distributed to homes, industry, schools, and other recipients and when water is drawn directly from a river, lake, or underground source.

10.5 RESEARCH STATUS ON WATER QUALITY

Influence of water regimes on nutrients availability in soils was established but how far this would affect the crop quality has value-added agricultural produces. Thanh and Biswas (1990) identified the parameters of water quality, which have adverse effect on agriculture and on industrial pollutants discharged into river system and pumping of brackish groundwater. Akanda et al. (2001) reported that ground and surface water have the differences in water quality within the parameter that are also influenced by the cultural practices (particularly use of chemicals) adopted for crop cultivation. Crop production is adversely affected by water-borne pollutants both physical and or dissolved chemicals that caused toxicities. Pollutant concentrations in irrigation water have been increasing which might be due to degradation of watersheds that replenish irrigation system. Onken and Hossner (1995) reported that the use of groundwater for irrigation has ingestion of crops under irrigation could be another exposure to arsenic and other heavy metal contaminations wherein there is every possibility of entering these toxins in food chain and might become disastrous. Paddy soils show arsenic levels elevated in the areas where arsenic in groundwater is used for irrigation was high and where the tube wells have been in operation for longer period of time. Thus, the mass awareness on water-saving technologies adopting microirrigation systems would create a favorable environment for sustainable agriculture.

10.6   IRRIGATION WATER QUALITY ON CROP GROWTH AND PRODUCTIVITY

The water quality parameters, such as pH, electrical conductivity, total dissolved salt and total soluble solids, alkalinity, nitrate, phosphate, sodium, and calcium, magnesium, can have a beneficial effect or adverse effect on irrigated agriculture. Impact assessment of adequate supply of usable quality water for the purpose of irrigation to influence the crop quality and marketability has a great deal to do with the present art of agriculture. A number of water management technologies are required to enhance agricultural production as well as profit. It was observed in agriculture where a gap occurs between the irrigation potential created and utilized in agriculture sector. The water management technologies developed so far were not also being practiced at the actual users end. Appropriate interventions are required to provide corrected measures to address the problems at the farmer's field as end users of the technologies on adoption of water efficient management practices for growing field crops. The influence of irrigation water quality on crop quality, marketability, and mitigation options of adverse effect of unsuitable contaminated polluted waste water on agricultural produces needs more attention at present day of precision agriculture to maximize the net returns. The mass awareness on water-saving technologies adopting microirrigation systems would create a favorable environment for sustainable agriculture.

10.7   WATER AND OTHER NATURAL RESOURCES

Soil and water are two basic natural resources that need effective conservation in efficient manner toward improving water productivity of crops. Because of this, rice coverage is about 75% of gross cropped area of the region. Out of the total rice growing area, at least 25% is coming under rice-rice system. There is an increase of about 250% in summer rice area in the state West Bengal during the last 22 years. Among the crop sequences followed by the farmers, rice-rice is most expensive for total water use. Increase in summer rice area with poor water management practices, overexploitation of groundwater under deep tube well irrigation system resulted in groundwater depletion at a faster rate having water table at below 9 m of surface during April–May is the main concern of the today's agricultural. In deep tube well command, groundwater is becoming rich in iron and toxicity of iron and imbalances of other nutrients particularly Zn, P, and K is emerging as a major nutritional problem for rice. Deposition of such iron as its oxides on oxidation ultimately is deteriorating the physical, chemical, and biological health of soil. More than 12 districts of West Bengal are facing arsenic problems. The fluoride threats are similar. The iron toxicity would have come as an outstanding problem in coming years, if adequate attention could not be drawn. The mechanization leads toward faster deteriorating of soil structure. As such, rice productivity level was almost static since last few years. The water-nutrient productivity of the rice-based system is to be optimized by improving soil environment. Organic fertilizers could play an important role in this direction to develop and provide balance nutrition for rice and its subsequent crops and quality produce in Indo-Gangetic basin.

With growing incomes, people express preferences for higher quality rice with preferred eating quality. The rice with short and roundish grain having low amylose content that becomes sticky after cooking, medium to long grain rice, with intermediate amylose and little aroma as well as parboiled long-grain with medium to high amylose content are mostly preferred. Market prices of rice are highly dependable on such quality parameters. The technical efficiency of fertilizer use could be increased in two ways:

1. The nutrient supply as per the need of the crop such as time of fertilizer application.
2. Application of organic fertilizers alone or in combination with inorganic fertilizers to improve efficiency. But application and effective prices of nutrients from organic sources are important.

The long-term adverse effects of inorganic fertilizer on soil properties and the environment could be ameliorated under integrated nutrient management using organic sources. Organic fertilizers are often seen as a means of sustaining long-term soil fertility and also as means of enhancing the efficiency of chemicals fertilizers, could improve soil characteristics and yield gains. The main goal of the study is to quantify the benefit of organic fertilizers of *kharif* rice under an integrated nutrient management program and reduce total water requirement of rice-based crop sequence, as a system approach. The influence of organic amendments on soil's physical, chemical, and microbiological properties on short- and long-term basis as well as its effect on quality parameters (like CHO, fat, and protein content along with parboiling and elongation including market price) have to be evaluated scientifically. This transformation might lead to increase bulk densities, penetration resistance, and sharing stress status of the soil. The organic fertilizers might have influence on the magnitude of porosity and may affect the size and rigidity of pores, which influenced root proliferation, gaseous exchange, and water retention as well as evapotranspiration (ET) from the crop field affects the water use of the crops directly.

The parameters of water quality, which have adversely affected irrigated agriculture, might have affected the crop quality. Ground and surface water have a same area in and within the parameter that is also influenced by the cultural practices adopted for crop cultivation. Rice productivity has reportedly been affected negatively by water-borne pollutants both physical and dissolved chemicals that cause toxicities. Irrigation water consists contaminants and have been increasing might be due to degradation of watersheds that replenish irrigation system and industrial pollutants discharged into river system or increased pumping of brackish groundwater. These factors drive into negative externalities and extract their impact on agricultural productivity. Studies on paddy soils show that arsenic levels were elevated in the areas where arsenic in groundwater used for irrigation was high and where the tube wells have been in operation for longer period of time. Use of groundwater for irrigation has ingestion of crops under irrigated agriculture has every possibility of entering these toxins in food chain and might become disastrous. They are emphasized for evaluation of field grown rice and vegetables which should be prioritized to examine the mode of accumulation of such chemicals in edible portion of agricultural produces affected by irrigation water quality.

Influence of water regimes on nutrients availability in soils was established but how far this would affect the crop quality is required to be assessed as an addition to get value-added agricultural produces. Nutrients in fertilizers can be leached and added to irrigation water being percolated in groundwater and be carried out by surface and subsurface drain. That also might affect the quality parameters of the crop and market prices of the produces. Thus, impact of water management on water quality from groundwater of different tube well sources could be assessed further. Surface as well as groundwater is used extensively for various purposes (drinking, agriculture, industrial, and navigation) and those became not suitable of being used for the purposes of irrigation to the crops due to chemicals as well as biological contaminations, which is also required to be assessed. Irrigation for crop cultivation and its subsequent effect on crop quality vis-à-vis effect of contaminated, polluted, waste water used for irrigation, and its subsequent effect on contents of such chemicals on edible portion of agricultural produces, its influence on market prices of the these commodities along with possible mitigation measures against adverse effect of such unsuitable quality of irrigation water has got importance.

10.8   INDUSTRIAL AND DOMESTIC USE

Water-soluble minerals, such as sulfate and carbonate precipitations are present in water heater. Hard water is softened to remove these ions of calcium and magnesium. The softening methods often substitutes sodium cations. Hard water makes suitable to soft water for human without any health hazards. This uses excess sodium and with calcium and magnesium deficiencies may cause health problems.

10.9   ENVIRONMENTAL USES

Water quality is an important parameter which denotes the environmental water quality, also called ambient water quality, its existence in all water bodies, such as stream, lakes, and oceans. Water quality standards and its maintenance for surface waters vary remarkably due to different environmental states, ecosystems, and intended human utilization. It is noticed that hazardous toxic substances are present in water, such as irrigation, swimming, fishing, boating, and industrial uses. They have immense bad impact on as a wild habitat across the world. Heavy metal substances on fresh drinking water many wild habitats get extinct around the world. Environmental scientists are concentrating on achieving goals for maintaining healthy ecosystems and may focus on the protection of populations of extinct species and ensuring better human health. Some biological monitoring metric has been scaled up in many countries; the aquatic insect orders, for example, Ephemeroptera (e.g., mayfly), Plecoptera (e.g., stonefly), and Trichoptera (e.g., caddisfly). There are bivalve molluscs which are largely used worldwide as bioindicators to monitor the health states of the marine ecosystems. Their population status, physiology, behavior, or the level of contamination with toxic hazardous heavy metals can indicates aquatic environment.

10.10   WATER QUALITY IN AGRICULTURE

Irrigation-dependent agriculture needs an adequate water supply of better quality. The water requirement in different sectors is increasing. Water is used in an intensive way which says that all kind of irrigation projects seeking supplemental supplies must rely on lower quality water. Supplied poor quality water with potential water quality-related problems decreases production. Quality of water refers to certain physical, chemical, and biological characteristics. Irrigation water quality, water characteristics of the other factors are considered to play a significant role. There are different quality needs and one water supply is considered more acceptable if it produces significant results than an alternative water supply. There is a sediment load, to remove the precipitate. In the same way, snowmelt water is more safe than other water for drinking purposes.

The crop used water and frequency of such use with contaminants heavy metals should be judicious. The different water qualities gained from observations that develop following use and have a cause-effect relationship between a water constituent and the water problem. The water quality-related problems which are essential for irrigated agriculture and some of the characteristics are then guidelines related to suitability of water or not. Such several guidelines have become available covering many types of irrigation water use and its quality parameters. Irrigated water utilized and quality depends upon the type and quantity of dissolved soluble salts. Irrigation water with salts is applied as water evaporates or is used by the crop. It is clear when there is no availability of good quality of irrigation water.

The fate of these salts is that it accumulates from prior irrigations and can be leached below the rooting depth. In salinity problem areas, leaching with irrigation water is needed. Year after year, salinity can remove this salt from the soil it needs to be leaching out carefully. The amount of salinity, leaching required depends upon the irrigation water quality to remove such salt from the soil. The lower rooting depth salinity depends upon the leaching that has occurred or affected root zone.

Under irrigated agriculture when irrigation is provided in long time gap to the crop by ideal irrigation, that is, surface methods and conventional irrigation method, then crop yield is reflected on root zone salinity; on the contrary for crops irrigated on a daily basis, crop yields affect the status of soil salinity. The crop significantly responses to the average root zone salinity contribution. Water-bearing strata lying between 2 m which places salinity problems to come up. Accumulation of salt in shallow water table is thus essential for salinity control in irrigated agriculture. Hazards of water table like drainage make long-term irrigated agriculture nearly impossible to gain without adequate drainage. Leaching of water is applied to check salts within the tolerance levels of the crop.

Water quality always depicts an important role on the soil surface, infiltrates too slowly to supply the crop with enough water to sustain acceptable yields of the crops. The infiltration rate of water into soil and bond of soil have also greatly influenced the intake rate of solution. These are the salinity present in water, also calcium and magnesium are present in the soil. Under these both factors may work at the same time, and secondary effects may be longer for an extended period of time. Some effects on fields are crusting, weeds, low-lying wet situation, nutrition and crop failure, and rotting of seeds. Many issues are present regarding after effects of an infiltration problem.

Water quality generally occurs in the surface few centimeters of soil and also infiltration creates a problem which is related to water quality. Therefore, it is linked to the structural bonding of this surface soil and its low calcium content related to that of sodium. Sometimes, a soil is irrigated with high sodium water and it causes soil structure imbalance. Close soil pores or capillary pores are present in the layer, upper layer of the soil. There is another physical problem caused by extremely low calcium content present on the surface soil. Under these circumstances, low salt concentration can cause a similar problem as it gets bond with soil, including calcium, from the surface soil. Toxicity problems occur if certain ion in the soil and water which directly and indirectly affect the crop yields. The crop losses may occur if even massive concentration and the crop sensitivity results in crop failure or drastic yield loss. The annual, perennial crops, and higher trees are more sensitive. The physiological changes may occur inside the plant. Under these situations, plants show its symptoms and it is first evidenced by marginal leaf burn and interveinal chlorosis and necrosis. Plants under high concentration of water salt also uptake more where reduced yields may occur. In such situation, it is also found that the more tolerant annual crops are not sensitive even less salt crops will be damaged or growth will be hampered if concentrations of such salts are significantly high for crossing the critical limits.

The salt concentration levels present in the soil makes an important issue to the plant. It is observed that the uptakes of ions are transported to the plant parts where the accumulation takes place. The salt concentration is high where the water loss is greatest through most of the leaf tips and leaf edges.

Accumulation of toxic salt causes damage to the crop plant. Such uptake form of ions, its effects on plants depend on salt concentration. Under arid climate or dry season, salt deposition is more and it might show little or no damage. The nitrogen dose if exceeds the requirement limits in the water may cause excessive vegetative growth and extend crop maturity. Irrigation with sprinkler spot or deposit of chemicals on fruits or leaves may occur with high bicarbonate water, gypsum, iron, and pH. Some precipitates also reduce further the water infiltration rate of slowly porous soil. The water quality may occur and soil status that may restrict its use or that of agronomic management to maintain desirable yields. The evaluation pattern consists of such issues of which one is water quality problem. The evaluations of issues are done in terms of use and the farm management techniques of the water user.

10.11   PROBLEMS ENCOUNTERED WITH WATER QUALITY

There are four problem categories, that is, salinity, infiltration, and toxicity and miscellaneous are used for evaluation. Water quality problems are seen of crop production problems. It is difficult it is to formulate an agronomic management for water quality solution if the problem is more complex and irrigation water quality factor is considered categorically. The guidelines formulate for number of factors are evaluated. These are discussed below:

a. Salt concentration causing the problem.
b. The soil-water-plant interactions.
c. The severity long-term use of the water.
d. The management to prevent and correct problem.
e. For irrigated agriculture, the quality of irrigation water should be standardized. These assumptions must be clearly known but should not become rigid prerequisites regarding quality. Alternative guidelines can be ready if actual conditions of use differ widely from those observed.
f. Some set of cropping patterns are experienced when using water with restrictions in selection of crop and management alternatives is required. The water is used when severe restrictions result in cropping problems and reduced yields. The water quality values depend on the economics of the farming and cropping techniques required.
g. The problem water, that is, wastewater, such as pesticides and organics where the good quality of water is always desirable. The fresh and pure water is always advisable for drinking water, live stock, salinity, and others bovines.
h. This is important as irrigation supplies are also commonly used for human drinking water. The WHO is consulted for water quality assurance.

10.12   REMEDIAL MEASURES FOR AMELIORATIONS

Best ameliorative is management practices for poor quality of water. Quality of water concerned, for crop production appropriate management practices facilitates higher yields. Some of the management practices are described below:

a. **Gypsum application**

Among the ameliorants, gypsum is one of the chemical amendments; when added to water will increase t h e calcium concentration in the water thus improving the infiltration rate of water. Gypsum requirement is estimated with relative concentration of Na, Mg, and cations in irrigation water. To add 1meq/low calcium, 860 kg of gypsum of 100% purity per ham of water is necessary.

b. **Irrigation management**

Irrigation is one of the management options for alleviating salt; some crops are susceptible at germination and growth stage with salinity, but it was observed that they are tolerant at later stage. Any kind of trace element and its toxic effect on plants may be revived with many practices to ensure good quality water.

c. **Fertilizer application**

Many soluble salts can be ameliorated by the fertilizers, manures, and soil amendments which include many soluble salts in high concentration if placed too closed to the germinating seedling or the growing plants. The fertilizer may cause or aggravate a salinity or toxicity problem. Application of fertilizer in placement swells as timing of fertilization. Application of fertilizer in little amount can revive plants from health hazards.

d. **Methods of irrigation**

Water use efficiency directly depends upon the method of irrigation applied. Sprinkler irrigation and its utilized water should not be of poor quality. Water with Na and Cl in sprinkler irrigation may damage the leaf and cause leaf burn.

e. **Crop plant tolerance**

The crop plants can affect the quality of water and its bad impact on crop plants. Specific metal causes plant effect which hampers plant growth and increases the yield.

f. **Sowing method**

Seed germination is hampered with significant level of salinity. Planting methods are adopted to avoid the germination failure. Seed should be placed in the area where there is less salt present in the soil. It was seen that lower portion of the ridge is low salt content in soil. So, seed placement is to be done in the slope of the ridge.

10.13 QUALITY OF IRRIGATION WATER

Saline soil develops from many factors including soil type, field slope and drainage, irrigation system type, management, and fertilizer system. There is another way to overcome the problems like soil and water management practices which are an important part. Factors affecting crop yield and soil's physical status, irrigation water quality can affect fertility requirements, irrigation system performance. Important management concerned is when to apply, how much apply, and where to apply irrigated water. Irrigation water should be good quality for the management of salt is required for long-run water productivity.

10.14 CRITERIA OF IRRIGATION WATER QUALITY

Criteria of irrigation water quality are categorized below:

- Soil and water salinity hazard.
- Soil and water sodium hazard is in relative proportion of sodium to calcium and magnesium ions.
- Soil pH (acidic or basic).
- Soil alkalinity (carbonate and bicarbonate).
- Specific ions, that is, chloride, sulfates, boron, and nitrate.

TABLE 10.1　Salinity Hazard of Irrigation Water Based Upon Conductivity.

Limitations for use	Electrical conductivity (dS/m)[a]
None	≤ 0.75
Low	0.76–1.5
Medium[b]	1.51–3.00
High[c]	≥ 3.00

[a]dS/m at 25°C = mmhos/cm; [b]Leaching at higher range; [c]Good drain and at germination.

The crop productivity depends on the water salinity hazard that is calculated by electrical conductivity (EC_w). The high electrical conductivity of water on crop productivity at initial stage makes water unavailable. Less water is available when higher electrical conductivity takes place in plants, even though the soil may appear wet but plants cannot uptake. The plants can only uptake pure water where usable plant water in the soil solution decreases drastically as electrical conductivity increases (Table 10.1).

The amount of water uptake by crop is directly related to crop yield. High electrical conductivity (EC_w) reduces yield potential and the crop yield.

Laboratories and literature sources reveal some problems parameters are salinity and total dissolved solids (TDS). These are dissolved salts, that is, ions and charged particles in a water sample. Hence, a TDS is a direct measurement of dissolved ions and electrical conductivity is an indirect measurement of ions by an electrode method. Generally, it assumes that and similarity of "salinity" with common table salt or sodium chloride (NaCl); electrical conductivity measures salinity from all the ions dissolved in a given sample. These are the negatively charged ions, that is, Cl^- and NO_3^- and positively charged ions, that is, Ca^{++} and Na^+. The preferred unit is deciSiemens per meter (dS/m) or millimhos per cm are used. Some conversion factors used to change between unit systems are provided in Table 10.2.

TABLE 10.2 Conversion Factors for Irrigation Water Quality Laboratory Reports.

Component	To convert	Multiply by	To obtain
Water nutrient or TDS	mg/L	1.0	ppm
Water salinity hazard	1 dS/m	1.0	1 mmho/cm
Water salinity hazard	1 mmho/cm	1000	1 μmho/cm
Water salinity hazard	EC_w (dS/m) for EC <5 dS/m	640	TDS (mg/L)
Water salinity hazard	EC_w (dS/m) for EC >5 dS/m	800	TDS (mg/L)
Water NO_3N, SO_4-S,B applied	ppm	0.23	lb per acre inch of water
Irrigation water	Acre inch	27,150	Gallons of water

i. Sodium in irrigation water

The plant growth is initially limited by the salinity level of the irrigation water; the utilization of water with a sodium imbalance can further reduce yield under certain soil texture conditions. The irrigation water present high sodium to the calcium and magnesium, this is termed as sodicity. The excessive soil accumulation of sodium in the form of sodic water is not the same as salt water. Soil sodicity shows the soil abnormality, such as soil swelling and dispersion of soil clays and sometimes sodicity shows surface crusting and pore clogging. In degraded soil condition, in turn problems of infiltration cause increased runoff of due to heavy soil erosion. Sodicity causes downward movement of water into the soil, and actively growing plant roots do not get adequate water.

Sodicity assessment present in water as well as in soil is called sodium adsorption ratio (SAR). The SAR refers to the relative concentration of sodium (Na) compared with the sum of calcium (Ca) and magnesium (Mg) ions. The SAR assesses the potential for infiltration hazards of sodium imbalance.

Mathematical equation of sodium absorption ratio is adjusted irrigation water with high bicarbonate (HCO_3) content, as SAR (SAR_{ADJ}) can be estimated. The probable soil infiltration and permeability problems may occur from the utilization of irrigation water with high sodicity cannot be enough assessed. The potential of swelling has low salinity water is greater than high EC_w waters.

$$SAR = \frac{Na + \frac{mg}{L}}{\sqrt{\frac{(ca+++)\left(Mg + + \frac{mg}{L}\right)}{2}}}$$

Soil's physical properties play an important role for balancing the sodium content in soil and factors, that is, soil texture, organic matter content, cropping system of the area, irrigation management system affects sodium. Soil reduced infiltration rate and crusting from water with elevated sodium absorption ratio if is more than six than those containing more than 30 percentage expansive smectite clay. Soils having more than 30% clay include most soils in the clay loam, silt clay loam type.

ii. Alkalinity and pH

Soil acidity or basicity depends upon irrigation water supplied, expressed in pH, if less than 7.0, it is acidic and if it is more than 7.0, it is basic. The normal pH of water is from 6.5, but for irrigation it may slightly change. Soil having high pH, more than 8.5 is often caused by high bicarbonate and carbonate concentrations known as alkalinity. Soil having high carbonates and its effect may be observed in soil-like formation of calcium and magnesium ions. The presence of alkaline water intensifies the impact of SAR and may cause serious problem. To overcome this problem, cleaning and other acidic materials into the system is essential.

iii. Chloride (Cl)

Among the microelements, chloride is essential to crop plants in very low amounts; its deficiency symptoms can cause toxicity to sensitive crops at high concentrations. High concentration of chloride results in more problems when applied with sprinkler irrigation. Problems like leaf burn under sprinkler from chloride can be reduced during night time irrigation to avoid this symptom. Sprinkler should not be in direct contact with leaf surfaces ground areas. There are some crops having low chloride tolerance like dry bean, onion, carrot, lettuce, pepper, corn, potato, and alfalfa, sudangrass, zucchini squash, wheat, sorghum, sugar beet, barley are high tolerance.

iv. **Boron (B)**
Among the micronutrients, boron is needed in low amounts. Sometimes, it shows toxicity at higher concentrations. Crop shows its toxicity even with less than 1.0 ppm concentration. Most of the soils contain enough boron. Boron toxicity occurs even at low concentrations, and additional boron may lead to toxicity. Boron in the irrigation water is approximately equal to these values or slightly less.

v. **Sulfur (S)**
It is important for crop's growth and development, particularly in oilseed crops for the synthesis of fat sulfate in irrigation water is beneficial, and irrigation water in most of the soils often has enough sulfates for maximum production for most crops, but additional application is required in oilseeds crops. Some exceptions occur in sandy soil with less than 1% organic matter and show less than 10 ppm sulfate in irrigation water.

vi. **Nitrogen (N)**

Among the primary nutrient of the plants, nitrogen is the major contributor and its deficiency shows crop loss, and nitrate nitrogen it is a common and available form that can be a significant N source. Irrigated water with high nitrogen results in nitrogen toxicity which occurs in barley and sugar beets. Hence, these problems can usually be overcome by good fertilizer and irrigation use efficiency. The crop and nitrate relationship toward the concentration should not exceed 10 ppm nitrate nitrogen.

10.15   MAJOR NUTRIENT GUIDELINES

a. The quantification is essential regarding the benefit of organic fertilizers on yield, growth, and productivity of rice under an integrated nutrient management schedule and reduced total water requirement of rice-based crop sequence.

b. The organic amendments play a major role on soil's physical, chemical, and microbiological properties; their effect on different aspect of quality parameters, that is, CHO, fat, and protein content is more.

   c. The organic nutrients play an important role on the porosity and results in the size and rigidity of pores, which influence root proliferation, gaseous exchange, and water retention, as well as ET from the crop field affects the water use pattern of the crop plants.

10.16 SUMMARY

The quality of irrigation water plays a significant role to farmers and its impact on plants, the crop productivity, water infiltration rate, and other soil physical status. Benefit of organic fertilizers are obtained on yield, growth, and productivity of the crops under an integrated nutrient management schedule that reduce total water requirement based crop sequence. The organic amendments play major role on soil physical, chemical, and microbiological properties, their effect on in different aspects of quality parameters, viz: carbohydrate, fat, and protein content is more pronounced. Focus is needed in integrated manner to overcome the toxic hazard and grain quality issue. These also play important role on the porosity of the soil, the size and rigidity of pores, which influenced root proliferation, gaseous exchange, and water retention as well as evapotranspiration (ET) from the crop field, they affect the moisture extraction pattern of the crop plants. The irrigation water and its source affect the soil plant that should be analyzed in a reputed laboratory to fix up the problem for proper remedial measures.

KEYWORDS

- **water quality**
- **impacts on agriculture**
- **research**
- **methodology**
- **crop productivity and quality of produces**
- **prices**

CHAPTER 11

River Linking Project

ABSTRACT

Water resources of the country totally depended on annual precipitation, where there was heavy downpour within shorter period, it caused inundation of the surrounding termed 'flood' that is either annual or alternate occurrence to many places, drained out with down streams in course of time, causing natural devastation in a locality. The pattern of rainfall is changing with global climate change. Uneven, erratic and uncertainty of rainfall aggravated the situations. So the "excess" water due to heavy downpour in shorter period might resemble with "surplus." There is no scientific basis to arrive at the conclusion that any river basin is "surplus" or "deficit", on the basis of such effect, since entire implementation options in any river basin. It is also essential to assess the environmental and ecological impacts due to the canal-link project including barrage and cross-drainage structures. The study includes assessing the environmental impacts related to the location, design, structure, and operation of the project. A benchmark survey to assess economic impacts and to quantify the influence on flora and fauna along with effect on biodiversity is required. The interlinking project concept states the availability of surplus runs in some deficit river basins.

11.1 INTRODUCTION

In India, developmental process has been initiated after independence in changing the production structure of the economy. The share of agriculture in the total gross domestic product (GDP) has got reshaped and remains the economy's major growth driver. Agriculture not only provides gainful employment and income, it also generates demand for nonagricultural goods and services. Slowdown in the growth of agriculture, thus, precipitates a drop in demand for industrial goods and services. Irrigation is one of the crucial inputs for achieving sustained agricultural growth and reducing inequality and poverty. After independence, significant progress has been made in the provision of irrigation facilities. The new agricultural technology is more water-intensive and in the absence of adequate and timely irrigation there will be no great productivity gains.

11.1   OBJECTIVE AND IMPORTANCE

The main aim of National River Linking Project is the transfer of water from excess water to deficit water basins. The transfer of water through inters basin water transfer projects. To the extent of further clarifications, into the term surplus states that it is the additional water and it meets the human's requirement for irrigation, consumption, and industries. Water resource available period caused inundation of the surrounding termed "flood" that is either annual or alternate occurrence to many places drained out with down streams in the course of time causing natural devastation in a locality. Uneven, erratic, and uncertainty of rainfall aggravated the situations. So the "excess" water due to heavy rainfall in shorter period might be resembled with "surplus." The river basin is surplus or deficit, on the basis of such effect, as entire implementation options in any river basin is dependent on it. The canal project includes barrage and cross-drainage structures. The study includes assessing the environmental impacts related to the location, design, structure, and operation of the project. The impact due to the project location, the resettlement and rehabilitation of displaced families, and assessment of loss of forest, natural reserves should be discussed. A benchmark survey to assess the economic impacts and also to quantify the influence on flora and fauna along with effect on biodiversity is required. Augmenting of irrigation and unemployment in certain pockets of India is a natural phenomenon. The effects that can damage the ecosystems were built through millions of years of evolution. There is land available to resettle the people evacuated in the process other than pressure on forests, pastures, and wetlands occurred.

The interlinking projects formulated has to be taken up with serious consideration that might not be an effective solution of diverting "surplus" water to "deficit" zone or to control flood other than coverage of more irrigation potential.

11.2 CONCEPT OF INTERBASIN WATER SUPPLY

The river interlinking project concept states the availability of surplus runs in some river basins. In the conceptual framework, one can simply utilize. The largest infrastructure project developed ever undertaken in the world, transferring water from river basins to facilitate the water acute shortages in drought prone zone of India. In contrary, north and eastern part of India face flood which sometimes exceed the water level; difficult to survive. The project proposes to build 30 links and approximately 3000 storages to facilitate the navigation of river water. The estimate of key project will cost a staggering Rs. 5,60,000 crore, handle 178 km³ of interbasin water transfer per year in hydropower capacity, add 35 million ha irrigated areas, and navigation and fishery benefits (Rath, 2003). Approximately, 3700 MW is required for lifting water. The average of two district supplies 33 km³ of water and the latter will transfer 141 km³ of river projects.

11.3 NATIONAL RIVER LINKING PROJECT COMPONENTS

There are two components under Interlinking River Project (ILRP). They are:

A) The Himalayan component

Himalayan component transfers 33 billion cubic meter of water through 16 river links. It has two sublinking. They are:

1. Ganga and Brahmaputra basins—Mahanadi river basin.
2. Eastern Ganga and Chambal and Sabarmati river basins.

B) The Peninsular component

The peninsular regions transfer 141 billion cubic meter of water with 14 river links. The peninsular region and its components have four subcomponent linking. They are given below:

1. Mahanadi and Godavari basins—Cauvery, Krishna, and Vaigai rivers.
2. West flowing rivers south of Tapi—north of Bombay.
3. Ken River, Parbati, Kalisindh river, and Chambal river.
4. West flowing rivers and east flowing rivers.

A) Himalayan component

The Himalayan component, with 16 river links, has two subcomponents: It includes the transfer of the excess water of the Ganga and Brahmaputra Rivers to the Mahanadi Basin. From there water is transferred to Godavari, Godavari to Krishna, Krishna to Pennar, and Pennar to the Cauvery basins. Second, the eastern Ganga tributaries water is transferred to irrigate the western parts of the Ganga and the Sabarmati river basins. This irrigation facility mitigates floods in the eastern parts of the Ganga basin, and provides the western parts of the basin with irrigation and water supplies. The Himalayan-based rivers have to conserve water and transfer flood waters from the tributaries of the Ganga and Brahmaputra and Godavari rivers.

B) Peninsular component

The Peninsular component has 16 major canals and 4 subcomponents. They are given below:

1. Linking of the Mahanadi-Godavari-Krishna-Cauvery-Vaigai rivers.
2. Linkage of west flowing rivers, that is, Tapi and north of Bombay.
3. Linking the Ken-Betwa and Parbati-Kalisindh-Chambal rivers.
4. Transfer of irrigation water to eastern part of India. The route irrigation under the Peninsular component is expected to irrigate a substantial area. Irrigation through this basin gets benefitted from the arid and semiarid western and Peninsular India. The estimated cost of the project of following three components is as follows:

    1. The Peninsular basin cost Rs.1,06,000 crore.
    2. The Himalayan component cost Rs.1,85,000 crore.
    3. The Hydroelectric component cost Rs. 2,69,000 crore.

The amounts of transfer water about 141 km^3 in peninsular zone. The Himalayan component covers about 33 km^3. The total power generated via the Hydroelectric component will be 34 GW in the Peninsular component (Rath, 2003).

The net water flows of the six interbasin water transfers projects are already operational in India, namely, Sharda-Sahayak; Beas-Sutlej; Madhopur-Beas Link; Kurnool Cudappa Cana; Periya Vegai Link; and Telgu Ganga. The river linking project proposes to build 30 river links and more than 3000 storages to connect 37 Himalayan and Peninsular rivers. The environmental concerns, rehabilitation, resettlement, socioeconomic conditions, and lack of alternative water management options. The major links connect the Himalayan Rivers and Peninsular Rivers. The river project consists of 30 river links and 3000 storage structures which supply 174 billion cubic meters of water (Smakhtin et al., 2007).

11.4 MAJOR RIVER BASINS AND WATER SURPLUS

The river linking projects are classified into three major river basins, that is, the Brahmaputra in the north east, the Mahanadi, and the Godavari basin in south. The peninsular regions cover about 6.5 km^3 of water (NWDA, 2012). The water transfer occurs from Brahmaputra to Ganga, Ganga to Subarnarekha, and Subarnarekha to Mahanadi under river linking project. The quantity of water which is excess flows after fulfilling the demand of agricultural, domestic, and industrial sectors and others needs.

11.5 ADVANTAGES OF RIVER LINKING PROJECTS

The following advantages are stated below:

a. Irrigate to 35 million ha of crop area and water supply to domestic and industrial sectors.
b. Adding 34 GW of hydropower potential to the national grid.
c. Mitigate floods in Eastern India.
d. Basin gives support to internal navigation, fisheries, groundwater recharge, and environmental flow of water. The per capita water storage capacity should be more than 200 m^3 per person, as compared with 5960, 4717, and 2486 m^3 per person available in the United States, Australia, and China, respectively.
e. Household water need is important in terms of first priority. The river basins have surplus water which can support to all water needs and agricultural purposes.

11.6 BENEFIT OF INTERLINKING OF RIVERS

The whole success of river projects under National Perspective Plan helps of 35 million ha area of irrigation, and provides 34,000 MW of hydro power. The other advantages are flood, navigation, water supply, and fisheries control.

11.7   SELF-SUFFICIENCY IN FOOD PRODUCTION

Cropping patterns and irrigation demands for achieving food self-sufficiency are key plans for the river linking projects. Three concerns dominate self-sufficiency assumption. Agriculture is the main source of healthy economic status. Foreign money improves nation and the estimates of food and water demand.

11.8   FOOD GRAIN DEMAND UNDER INCREASING POPULATION

About 45 million tons of foodgrains will be wasted by 2050. There are some declining trends of food grains per capita since 1990. Food grain production is decreasing and vegetable and animal products in the diet are increasing. The food grains demand could be significantly higher in India and the need for food grain may increase about 380 million tons by 2050. The feed grain demand of 120 million tons is almost 10-fold increase from the present levels. Due to the reduction of irrigation, supply for such a difference of food grains is justified in variability and near about 30% irrigation options should be there for achieving the target.

11.9   RURAL LIVELIHOOD AND EMPLOYMENT

Rural employment depends on the cropping pattern which is directly related with the supply of irrigation water. If there is enough supply of irrigation throughout the year, crop diversification and employment generation is also significant. Indian youth are mostly engaged with their farming for livelihood. Indian villages have different perceptions and priorities and way of thinking is changing for income generation. Indian youth always wants more income for their family even farming is common, but mostly remunerative in different industrial sectors.

11.10   SOCIAL COST-EFFECTIVENESS

The social cost-effectiveness is described below:

- Direct and indirect benefits of irrigation water transfers.
- Groundwater externalities of surface water transfers.

- Gender impacts and equity issues of new water transfers.
- Benefits of domestic and industrial water transfers.
- Environmental benefits and harmful effect.
- Hydrological feasibility.

11.11 SOCIOECONOMIC AND ENVIRONMENTAL ROLE

It is essential to assess the environmental and ecological impacts due to the link canal project including barrage and cross-drainage structures. The study included assessing the environmental impacts related to the location, design, structure, and operation of the project. The impacts due to project location should be assessed on the basis of resettlement and rehabilitation of displaced families, assessment of loss of forest, natural reserves. Detailed benchmarks survey to assess economic impacts and also to quantify the influence on flora and fauna along with the effect on biodiversity is being carried out. Interlinking of rivers is advocated as a solution to flood control, water deficit, augmenting irrigation, and unemployment in certain pockets of India (Zaman, 2004). The argument that water will be taken from regions with floods to regions with drought does not hold true. Smakhtin and Anputhas (2006) reported the effects that can damage the ecosystems were built through millions of years of evolution. The land available to resettle the people evacuated in the process other than pressure on forests, pastures, and wetlands. The effective way of diverting "surplus" water to "deficit" zone or to control floods other than coverage of more irrigation potential, only which could be obtainable from several alternatives, such as decentralized water-shed development, rainwater harvesting, groundwater recharge, receiving the existing local systems of water harvesting (Zaman and siddque, 2004).

11.12 REVIEWS ON SOCIOECONOMIC AND ENVIRONMENTAL IMPACTS

India is a unique country where one can see flood and drought coexisting in different parts of the country (Zaman, 2003). The unique geographical location of the country which makes some regions of the country water surplus while other water scarce. The idea of ILR was mooted to balance out this unevenness in the distribution of water. A number of teams manned by leading geologists and scientists from world acclaimed institutions have worked on the feasibility prospect of this project since 1960s. But nothing has taken a concrete form yet, because of the reason that depending on the scale of the project it becomes very tough (Amarasighe et al., 2007). The reasons of adoption of such projects along with its merits and demerits are discussed.

The reasons illustrated as therein on:

1. Indian monsoon sustains from July to October; the amount of rainfall in southern and western part is low. The river linking project helps these areas to supply water throughout the year as irrigation and many other purposes.
2. Indian farmers are very weak economically to run their family (Falkemark et al., 1989). River project helps to store water can be transferred from water surplus area to deficit zone.
3. The Ganga and Brahmaputra basin floods every year. The water diversion is required where there is scarcity of water. This can be achieved by linking the rivers. The flood will be controlled and scarcity of water will be reduced.
4. The river projects have commercial importance on a longer run. Many objectives which help in faster movement of goods from one area to other.
5. River projects enrich the aquaculture where people abundantly catch the fish which secure source of income.

11.13   ADVANTAGES OF RIVER LINKING PROJECTS

i.   **Irrigation:** River basin helps land areas which will not otherwise be irrigated and are wasteland and become fertile.

ii.   **Flood prevention:** River projects have well water drainage channeling system to drain out excess water from the surplus water zone.

iii. **Generation of electricity:** There is an additional advantage of storing large quantity of water with the same effort; electricity can be generated easily.

iv.   **Navigation:** River project facilitates the navigation system and system developing under this project; and additional benefits can be obtained from the same projects.

v.   **Potable water:** Drinking water supply might be facilitated to the areas of scarce drinking water.

11.14   DISADVANTAGES OF RIVER LINKING PROJECTS

i.   Ecological submergences of forests and wildlife might cause one of the major concerns that rivers could change their course in 70–100 years.

ii.   Initial cost is too high of the project.

iii.   Geologically, it would be tough to connect northern rivers with southern rivers due to slope.

iv.   The future project may not benefit as much as expected.

v.   The river linking projects will have invited political water sharing disputes along with other possibilities of harmful effect as follows:

1.   River basin has massive deforestation. With this activity, there will be indirect impact on rains and in turn, it will affect the whole cycle of life.

2.   Generally, rivers change their direction in about 100 years and it creates another problem on human and animal system.

3.   This project has fish production of fresh water which ends up to the marine ecosystem.

4.   The river project creates pond water reservoirs of these displaced people.

5.   The huge amount of money expends burden on the government and country will face debt trap from the foreign country.

11.15   CONCLUSION

The river projects work on diverse areas that might not be the effective solution of diverting "surplus" water to "deficit" zone or to control floods other than coverage of more irrigation potential only which could be obtainable from several alternatives, such as decentralized watershed development, rainwater harvesting, and groundwater recharge. Hence, in a nut shell, it can be concluded that ILR at all India level may not be a good choice. If possible, river linking can be done regionally to optimize the benefits while reducing the demerits.

KEYWORDS

- importance
- inter river linking project
- concepts
- components
- advantages

CHAPTER 12

Water Pricing

ABSTRACT

The water price is levied for its supply from a public or a private system with a view to ensure equitable water distribution, efficiency of the irrigation system, and its management. Policy on water price has been attracting attention of planners, policymakers, and researchers in view of its important role in regulating the water use within the reach. The declining per capita availability of water for diverse uses has generated serious concern in public domain. The Government's concern for a rational and pragmatic approach for levying water prices, in return for the water supplied to the users, made possible by the development and construction of new projects that requires huge public investment. Importance and necessity of levying water charges are now recognized for keeping adequacy in meeting the expenditure toward operation and maintenance for ensuring equitable distribution and its efficient use.

12.1 INTRODUCTION

Water tariff functions of users in centralized or semi-centralized systems for the suitable treatment, purification, and distribution of fresh, and therefore the sequent assortment, treatment, and discharge of waste product. Water evaluation observes vital economic conditions for social economic stability. Tariff setting practices vary wide round the world. During this factsheet, mounted water charges are delineate as a kind of tariff that is straight forward to water metering system.

12.2 GOVERNMENTS POLICY FORMULATION

Governments advise a range of policies and rules for mitigation of agricultural water management issues. So as to know whether or not these ways enhance property development; this thematic space evaluates:

a. The role of restrictive mechanisms in policy compliance.
b. The impacts of water laws gone totally to state governments if needed for best management of water as common pool resource. Multiple use of water ends up in higher water productivity in addition as water conflicts. Water as common property has been a key of competition for various agricultural production system strategies. The multiple use of agricultural water analysis examines:

   i. Contributions of common property water resource to livelihoods, notably of vulnerable groups.
   ii. Scale and institutional problems in managing trade-offs related to common property resource use.
   iii. Policies for best management of multiple use water resource, access, profit sharing, and maintenance of water resources. Variety of water establishments is functioning across the states and area unit effective in several ways in which the water establishment analysis focuses on the effectualness of water establishments in higher management, equity, adequacy, and timeliness in water convenience, programs, governance, and policies that support water establishments. There is currently agreement at varied levels that water is scarce and desires to be treated as an economic sensible. This rating is taken into account to be crucial for economic sensible is a lot sophisticated then different merchandise and services. This can be primarily thanks to its public administration across the countries. The assorted issues, viz., unskillfulness, poor designing, and social control, etc., associated with public administration have usually over up treating water as a social sensible rather than associating in nursing economic sensible. In fact, it is argued that it is "willingness to charge" for water and blocking price reforms (DFID, 1999).

These areas are unit public satisfactoriness, political acceptability, simplicity and transparency, Internet revenue stability, and simple functioned (Bolland and Whittington, 2000). Political and public interference clashes the basic objectives of pricing water. The price of water is equal to marginal cost for accessibility to end users. The setting of high tariffs is not acceptable to the public and politicians. The fixation of pricing mechanisms can be categorized into three, such as volumetric pricing, non-volumetric pricing, and market-based pricing (Johansson, 2000). In meter evaluation, water use is measured and charged. Formal water markets require tradable property rights in water that are conspicuous by their absence in most other countries. Formal water markets need tradable property rights in water that are conspicuous by their absence in most alternative countries. Water rights additionally facilitate cut back poorness and water conservation (Burns and Meinzen-Dick, 2005). The available water is not enough to meet the needs of users. Water markets seem to shrink of power regulation for groundwater extraction (Shah and Verma, 2008). The marginal value specification applicable that is acceptable as suitable or average value specification appropriate for estimating the water demand operations is being debated through empirical observation. The water price stable marginal prices, while some used both in order to imbalance the pricing in the water demand. Introduction of uniform pricing with rebate (UPR), which is capable of achieving the benefits of increasing block rate tariffs without adopting a block tariff structure being of utmost importance in agriculture, and the popularity of increasing block rate tariffs is attributed to the water professional's ignorance (Boland and Whittington, 2000).

There are huge variations in water rate structures and rate per unit volume of water consumed varies greatly for crops. The rate of water used for the same crop depends on season and type of system. Recovery of cost irrigation to pay is based on gross earning or net benefit of irrigation, water requirement of crops. Water cess is also imposed in some states of India. Irrigation water pricing is adopted and it varies within each country and it depends on topography of the land. A water price is fixed as per local conditions and costs of production. Irrigation water rates are desirable and volumetric efficient pricing mechanisms of water is under priced in most of the countries. As a result water prices neither reflect its scarcity value nor allocate efficiently. Similarly, no water rate is levied for agricultural purposes in most of the northeastern states except for Manipur. In Odisha, water rate is charged on flat basis for the staple crop of paddy under the command area of major and medium projects; the rate of water is varied under different types of crops.

To increase agricultural production, water pricing is important for marginal farmers. It will facilitate the end users to develop minor irrigation schemes, community development, operation, and maintenance.

Key performance indicators of such study are as follows:

- To enhance the agricultural yield.

- Creation of operational water for users based on the majority of farmers benefitted.
- Agricultural resources created by farmers to manage, operate, and maintain the projects.
- Fixing water prices is an urgent need at all delivery places and command areas and for its judicious use.
- To demonstrate 2–3 proven water-saving technologies in *Rabi* and summer (pre-*Kharif*) season taking 2–3 crops.
- To train the farmers for effective utilization of water resources for scaling up water productivity and agriculture sustainability.
- To disseminate the proven water management technologies at a faster rate to minimize gap between technology generation and adoption at end user's level.

12.3   ON-FARM PROVEN TECHNOLOGY ADOPTION

On-farm cost-effective technology is used for increasing water use efficiency. The different ways of application losses can be reduced through conveyance. According to crop type, water requirements also vary and are an important role for enhancement of water use efficiency. Irrigation water utilization and conservation and distribution are the basic parameters of on-farm water management. For better use and judicious use of water, optimum scheduling of irrigation, suitable method adoption, and conjunctive use of rain is adopted. Besides this, surface and groundwater for crop production with guide of drainage are perfect for optimum water management. Trist (1980) showed the irrigation methods for better irrigation as socioeconomic inputs. On-farm water management experiment was conducted as a collaborative project of Indian Council of Agricultural Research under the aegis of AICRP on Water Management (BCKV) in the deep tube well command of Nadia district in West Bengal for developing either package interventions or conducting component technology for field crops. The main aims of the study are given below:

i.   The site-specific and cost-effective water management technologies are suitable methods and optimum scheduling of irrigation for various field crops. The conjunctive use of rain and groundwater for crop cultivation and water economy. System develops for drainage in monsoon season to restore the soil workability condition for sustainable crop production. The water management technologies in farming, site-specific technology for poor farmers are for adoption with a view to exploit the valuable water resources. The technologies that are found suited by the farmers in their fields are as follows:

a. Raised bed and furrow method of irrigation for tomato, cabbage, and cauliflower.
b. Use of polythene lining in the field channel to reduce the conveyance losses of irrigation water.
c. Application of low depth of irrigation up to or less than 4 cm in diversified crops.
d. Intermittent application of water for summer rice with appropriate scheduling of irrigation reduced water requirement.
e. Wheat and mustard crops were introduced instead of continuum rice with proper scheduling of irrigation.
f. Rotation of water supply was made spout-wise to irrigate rice and wheat was grown simultaneously.
g. Adoption of minimal tillage in wheat and mustard.
h. Endorsement of low-cost channel lining materials with burnt clay, bamboo reinforcement concrete slabs.
i. Improved irrigation scheduling and methods were intervened with farmers' practices.
j. Adoption of border strip method of irrigation in wheat, mustard, and jute were introduced.
k. Improvement in conveyance losses of irrigation water through regular cleaning and maintenance of the earthen channels.
l. Compartmentalization of undulated/sloppy lands specially in rice field for smooth application of irrigation water.
m. Infrastructural intervention for smooth delivery of water through the existing spouts.
n. Growing intercrops in wider spaces of the main crops with water technology intervention.
o. Implication of drainage as an essential component in the monsoon and post monsoon seasons in the crop cultivation program.

12.4 ADOPTED TECHNOLOGIES AND ITS FINDINGS

i. The outcome of the proposed study would come to reduce the gap between irrigation potential created and its effective utilization minimizing wastage, reducing gap between technology generation and its adoption by end users in the field of ecology, health, water, and sanitation. The available technologies for its adoption after refinement promoting and making available safe and quality drinking water and water for irrigation for sustainable crop cultivation and production in different agroclimatic zones protecting against the disastrous effect environmental degradation and climate change under scientific calendar.

ii.   Zero tillage and raised bed planting could be adopted by the farmers for efficient moisture use. Drum seeding of crops during post rainy season might be producing encouraging results on and may be a viable alternative in rice-wheat crop sequence where labor is short supply and costly. Beneficial effects of conservation tillage with crop residues/mulching or green manure for improving moisture availability, controlling weeds, and regulating soil temperature are possible. Zero tillage was also found feasible in *Rabi* crops, following *Kharif* crops. Conservation agriculture, that is, tillage practices facilitated with mulching for enhancing water and nutrient use efficiency.

iii. The major output of this study would be established guidelines and recommendations for facilitating or accelerating expansion of effective utilization of lowland ecosystem and developing appropriate farm technologies for multiple uses to alleviate poverty and improve water and land productivity in the eastern Indo-Gangetic Plains or extrapolation anywhere is deemed to be appropriate. The study will provide policymakers and other stakeholders with knowledge on the benefits and cost-effective low land utilization and multiple uses of water technologies as well as the physical, institutional, and policy conditions that have led to rapid adoption in some areas but not in others.

Output 1: A cross-site comparison of the trends in adoption of appropriate technologies and in biophysical conditions, institutions, and policies that have either constrained or accelerated the adoption. The factors, such as the physical nature of the aquifer, access to credit, inputs, and tariff barriers to technology imports.

Output 2: Assessment of strategies and policy options for management of such wetland or *khal, jheels, beels* in conjunction with surface and groundwater resources for increasing productivity of land, water, and crops in such areas would generate sufficient amount of field data made available to the policy makers. In selected sites in the study area, this would take recharge and environmental externalities.

Output 3: An analysis of the trends in technology generation and adoption, extrapolation, revalidation has been conditioned by different conditions and policies across the four sites.

Output 4: Capacity building of national institutions for analytical policy research would be made possible. This will have a number of components including training of site staff and graduate students in data collection and analysis and in report writing.

12.5 WATER TARIFF FIXING

Water rates are different in some states like Tamil Nadu (1962) and Kerala (1974). Water rate is levied for agricultural purposes in northeastern states except for Manipur. The water from minor systems is supplied only on prepayment basis in West Bengal. In Jammu and Kashmir, Haryana, West Bengal, and Kerala variations in water rates appear to be marginal. In most of the southern states of Andhra Pradesh (AP), Tamil Nadu (TN), Karnataka, and Pondicherry water rates are fixed with land revenue and land is assessed as wet and dry. The difference between the dry and wet assessment is taken as water rate. In most of the states, public water supplies for irrigation are levied on the basis of area irrigated, while water charges are levied on the basis of the number of hours of watering or volume of water. In India, maximum working expenses in Gujarat (Rs. 4768 rupees/ha) followed by Maharashtra (Rs. 3050 rupees/ha) and minimum working expense is in Odisha (Rs. 256 rupees/ha).

12.6 AFFORDABILITY

Technology development of irrigation water covered under water-saving technologies is very minimal. It is costly for stored water tariff structure and more and more area is being brought under these scarcity conditions. Micro irrigation is important; among the irrigation, water-saving technologies are sprinkler and drip irrigation techniques. Sprinkler and drip irrigation techniques are covered to a diversity of crops. Mostly, horticultural crops are suitable for this and leading state is Gujarat; farmers use microirrigation systems on various crops, such as wheat, bajra, maize, groundnut, cotton, castor, and vegetables also (Kumar et al., 2004). In Maharashtra, drip systems are used even on water intensive (Narayanamoorthy, 2006). Micro irrigation saves 48–67% water, 44–67% energy, and 29–60% labor. Moreover, the farmers are interested in investing their money for irrigation without any subsidy (Kumar et al., 2004). Cost-benefits ratio analysis of drip irrigation in Maharashtra revealed (Narayanamoorthy, 2006) high return.

Efficient use of water is reported from many perspectives viz., riparian, federalist, formal law, civil society, environment, and economic (Iyer, 2003).

Water rights are given on an individual basis on 5 years contractual basis and no ownership of water is allowed. Optimal use of water is needed for achieving equitable and sustainable economic and social development. The pricing is achieving efficient allocation of irrigation water. These are proper valuation of water resources; institutional mechanisms like water farmers to support implementations of pricing policies.

12.7 FEASIBILITY OF WATER PRICING

Framework of price policies and tariff structures should take the following considerations into account in order to make the reforms feasible. These notes are public acceptability, political acceptability, simplicity, and transparency (Bolland and Whittington, 2000). The basic objective of water pricing is proper better allocation of water to the end users. For instance, for achieving economically efficient water allocation that gives highest return often, this results in the setting of high tariff structures, which are acceptable neither to the public nor to the policy maker. The volumetric water pricing based on acreage is found to be the easiest way of implementation administratively.

Water rates were revised only once during 1999 in Rajasthan. In Odisha, irrigation projects like *Pani Panchayats* started in 2002 and a total of 801 thousand hectares of irrigated command areas have been handed over to 13,284 *Pani Panchayats*. The types of farmers and the landless farmers are helped under *Pani Panchayat*. The people rebelled against the program and the model has collapsed (Das, 2006).

The power supply rationing can indeed act as a powerful tool for groundwater demand management (Shah and Verma, 2008). In the same way, charging of the power supply has shrunk the water use.

12.8 CONCLUSION

Suitable water management technologies adopted and disseminated to the farmers for enhancing agricultural productivity in the country with standardization of fixing water prices in agricultural uses. The water rates for surface water meter say this report by CWC. The pricing of water is levied for supply of water from a public or a private system. The main aim is to ensure equal distribution, efficiency, and its management. The important role is to regulate the water use within the reach and resources of the end users. The importance and necessity is standardized of levying water charges to the users and its adequacy in meeting the expenditure.

KEYWORDS

- **water price**
- **policy formulation**
- **technology adoption**
- **equitable distribution**
- **tariff fixing**

CHAPTER 13

Wetland Management in Water Productivity

ABSTRACT

In India, wetlands occur in all climatic zones and at all altitudes. Wetlands exist at below the sea level to about 6000 m elevation in the Himalaya. Wetlands occur wherever water accumulates for enough long periods that allow the establishment of plants and animals adapted to the aquatic environment. Water need not be present permanently and the depth of water stagnation may generally fluctuate. Periodically inundated wetlands square measure terribly effective in storing rain and have innate capability to recharge the bottom waters. Undergroundwater recharge happens through mineral soils found primarily round the edges of wetlands. The extent of groundwater recharge depends on the sort of soil and its permeability, vegetation, sediment accumulation within the lake-bed, extent to volume magnitude relation, groundwater level gradient, and period of water stagnation. Wetlands have an incredible ability to fulfill the water demand within the surrounding areas.

13.1 INTRODUCTION

Wetlands are an important ecosystem valuable for its diversity and habitat of thousands of aquatic flora and fauna. Wetland ecosystems are usually located at the ecotones between dry terrestrial and permanent aquatic ecotones. Under these systems, the water table is usually at or near the ground surface or the land is covered by shallow depth of water.

The oldest definition of the wetland was given by US Fish and Wildlife Service in the year 1956. Wetlands are the lands transitional between terrestrial and aquatic systems, hence water table is at or near the surface or land is covered with shallow water. Wetland also refers to encompass a diverse and heterogeneous gathering of habitats ranging from shallow rivers, floodplains, and rainfall lakes to mangrove swamps and salt marshes.

The convention on wetlands of international importance especially as water flow habitat is often known as the Ramsar Convention from its place of adaptation in Iran, 1971 is an intergovernmental treaty, which provides the framework for international cooperation for the conservation of wetlands. The Ramsar Conservation on Wetland in 1971, in Iran characterized wetlands as "areas of submerged and saturated soils, natural and artificial, permanent and temporary, freshwater or marine, low tide does not exceed 6 m." India has 19 wetland areas listed as Ramsar Sites and these are Chilika Lake, Odisha, Keoladeo National Park, Rajasthan, Wular Lake, Jammu and Kashmir, Harike Lake, Punjab, Loktak Lake, Manipur, Sambhar Lake, Rajasthan, Kanjli, Punjab, Ropar, Punjab, Ashtamudi Wetland, Kerala, Bhitarkanika Mangroves, Odisha, Bhoj Wetland, Madhya Pradesh, Deepor Beel, Assam, East Calcutta Wetlands, West Bengal, Kolleru Lake, Andhra Pradesh, Point Calimere Wildlife and Bird Sanctuary, Tamil Nadu, Pong Dam Lake, Himachal Pradesh, Sasthamkotta Lake, Kerala, Tsomoriri, Jammu and Kashmir, and Vembanad-Kol Wetland, Kerala.

The important components and features of wetlands are as follows:

a. They are distinguished by the presence of water.
b. They have soils which are not similar to those in adjacent uplands.
c. The vegetation ranges from partially flood tolerant to hydrophytic plants.
d. They are usually found between deepwater and terrestrial uplands with both the systems exerting their influences.
e. They vary widely in size, from a few hectares to several square kilometers.
f. They are found both inland and in coastal regions.

Some useful information has been collected during the last decade (Rai and Munshi, 1982) on wetland ecology and farming. Some work was done in the past on "the synergistic effects of hydrology, chemical inputs, and climatic conditions on wetland productivity and on how plants and animals adapt to stressful situations in various wetlands types has provoked ideas for further research on wetlands where there is need of integration of several disciplines."

Wetlands are known by various terms, indicating specific kinds of ecosystems with distinct vegetation, animals, and other characteristics: swamp, marsh, bog, fen peat land, mire, moor, muskeg, bottomland, wet prairie, reed swamp, wet meadow, slough, pothole plaza, hoar, bao, rbeel, diara, chaur, etc. These terms are decidedly regional or at least continental. The European classification is based on the amount of surface water and nutrient in flow type of vegetation, pH, and peat building characteristic. So, the physical and biotic characteristics grade continuously and the same term refers to vastly different ms on different regions. Based on geomorphic situation, topographical condition, textual class of the soil, water retentive capacity, and water quality including chemical nature, wetlands can be broadly classified into two major groups, viz.:

i. Inland wetland and
ii. Coastal wetland.

Floodplains and inundated paddy fields may be considered as seasonal wetlands. The importance of wetlands lies in their value as wildlife protection areas as well as areas of the water management and conservation for agricultural development. The highly fertile marshes and swamps, rivers and estuaries, and coastal areas of the lowland tropical region are used by man for many purposes. Wetlands are certainly important for many of the functions including regulations, production, carrier, and recreation.

The various useful functions of wetlands like sustaining life processes, water storage protection from storms and floods, recharge of undergroundwater, water purification, and depository for nutrients, erosion management of native climate, facilitate maintaining the ecological balance.

13.2   FLOOD WATER STORAGE

In their natural condition, most wetlands store flood waters briefly, protective downstream areas from flooding. By checking the floods, they maintain a relentless flow regime downstream, preserve water quality, and increase the biological productivity of the aquatic communities. This performance becomes more and more necessary in urban areas, wherever biological process activities, such as breaching of wetlands for residential, commercial, and industrial activities, paving of surfaces in structure areas, etc. have increased the speed and volume of surface water runoff and also the potential for flood harm. This necessitates the protection of wetlands, a very important means of minimizing flood damages within the future.

13.3   GROUNDWATER RECHARGE

Periodically, inundated wetlands square measure terribly effective in storing rain and have innate capability to recharge the bottom waters. Underground-water recharge happens through mineral soils found primarily round the edges of wetlands. The extent of groundwater recharge depends on the sort of soil and its permeableness, vegetation, sediment accumulation within the lakebed, extent to volume magnitude relation, and groundwater level gradient. Wetlands have an incredible ability to fulfill the water demand within the encompassing areas. Natural wetlands square measure underlain by aquifers with a high potential for water.

13.4   SHORELINE STABILIZATION AND EROSION CONTROL

Wetland vegetation will scale back bound erosion in many ways as well as:

   a.   Increasing sturdiness of the sediment through binding with stilt.
   b.   Moistening waves through friction.
   c.   Reducing current rate through friction, rising water quality.

Coastal wetlands significantly mangroves facilitate in bound stabilization and storm protection by dissipating the force by reducing the harm of wind and wave action.

13.5   WATER QUALITY

Wetlands play a vital role in raising the water quality by filtering sediments and nutrients from surface water. Aquatic vegetation helps in removing 90% of the dissolved nutrients like gas and phosphorus and conjointly in surface assimilation of serious metals. Dissolved materials could also be maintained in wetlands and therefore the water quality might vary seasonally or from year to year. Removal of sediment load is additionally valuable as a result sediments usually transport absorbed nutrients, pesticides, serious metals, and different toxins that begrime the water.

13.6   NUTRIENT RECYCLING

Wetlands, transition zones between land and water area unit are economical in filtering sediments. They will intercept runoff from land before it reaches the water and facilitate in filtering nutrients, wastes, and sediments from floodwaters. In sure wetlands, plants area unit is thus economical in removing wastes that artificial waste matter treatment systems use aquatic plants for the removal of pollutants from water. Wetlands take away nutrients, particulates, and total biological atomic number 8 demands from flooding waters for plant growth and facilitate over enrichment of different varieties of natural waters. However, overloading a soil with nutrients, on the far side its threshold impairs its ability to perform basic functions.

13.7 ECOLOGICAL BENEFITS

Wetlands being one among the foremost biologically productive natural ecosystems area unit are very important for the survival of various flora and fauna, as well as several vulnerable and species by providing shelter, food, etc. and form an area of the advanced food web. It is calculable that regarding two-hundredth of the far-famed species of life swear directly or indirectly on wetlands for his or her survival, as they are their primary and necessary seasonal habitats.

13.8 WETLAND PRODUCTION SYSTEM

Wetland production are fish, timber, reeds, medicinal and aromatic plants, fertile land, arable land, pastoral, fuel wood, transport, recreation, and tourism. By supporting various human activities, massive wetlands play a very necessary role within the subsistence and development of thousands of individuals.

However, makes an attempt to find out energy are created to spotlight on wetlands, giving stress on broad aspects of their situational points and correct utilization for the betterment of the world. The functions of a soil and therefore the values of those functions to human society rely on a complex set of relationships between the soil and therefore the different ecosystems within the watershed. A watershed may be a geographic region during which water, sediments, and dissolved materials drain from higher elevations to a typical low-lying outlet or basin a degree on a bigger stream, lake, underlying geological formation, or estuary.

13.9 CONSERVATION AND MANAGEMENT OF WETLANDS

There square measures monumental importance of conservation and management of wetland enhancing water productivity and agricultural property for up livelihoods in the Indo Gangetic basin. The aquatic scheme sometimes refers to property management of soil or lowland. Asian nation possesses 17.2 million ha of low-lying space out of 38 million ha within the world.

India, Bangladesh, and Burma are largely as semiaquatic crops whereby development stands for aquatic scheme product and services as major contribution to boost water policy. A mix of crop cultivation and farming together with fish culture has been steered (Thakur and Thakur, 1991) for such land state of affairs to extend the chance of employment moreover on offer macromolecule and fat within the dietary program.

The wetlands square measure the interphase between land terrestrial and permanent aquatic system and these square measure the saucer formed wet bodies, where water stands naturally a minimum for an inexpensive part of the year, is also of permanent, semipermanent, and temporary in nature, relying upon depth and sturdiness of submerging. Soil includes 6.4% (8.558 million km³) of the world's total area, of that 23.5 million ha coated in Asian nation, that accounts 25–30% throughout its peak wet periods (Anonymous, 1986), dominated largely in northeastern part of the country together with coastal areas. Actually, a similar environmental condition additionally exists in Southeast Asian countries like China, Japan, Philippines, Thailand, Malaysia, Sri Lanka, Bangladesh, etc. Survival of human civilization is being inextricably coupled with the wetlands since 4500 B.C. that James (1995) justified and termed the wetlands as "nature's kidney." The Ganges River, Padma, Mahanada, Mahanadi largely beneath Indo-Gangetic basin, and it is such a large amount of tributaries, canals, low-lying areas subject to frequent water work and flood creating them marshy and nearly turns out of traditional cultivation throughout the Kharif season. Such lands could ordinarily turn to agriculturally feasible solely throughout winter and summer season. Moreover, farmer's square measure victimization of these lands. Not consistently, however, in scattered approach, these marshy, fertile, productive wetlands square measure utilized sometimes by the associated farmers who are associated with the production of nutritious, and fashionable aquatic food crops, that is, deepwater rice, water chestnut or singhara, makhana or fox nut, or mythical monster nut and Colocasia spp., aquatic food decorative plants, that is, lotus, water lily, Royal water-lily; aquatic nonfood industrial crops, that is, Cyperus spp., hogla, shitalpati used for matreed; Calamus or rattan palms or bet, shoal, etc. used for creating ornamental unit materials; aquatic fodder crops; a large amount of aquatic helpful medicinal plants and aquatic weeds plants of biomass are also known. Besides, the introduction of fish in conjunction with this method may be a natural prevalence, the overall production of the system and economic production is not in the least encouraging thanks to technical copy and monetary stringency of the farmers. Indeed, this fertile and productive soil of wetlands will effectively be utilized following correct agro-techniques for sustaining higher production, which can flip economically viable for the upliftment of rural sector reducing financial condition and up livelihoods of the folks within the country. Through case studies on water chestnut and makhana in "Tal" wetlands beneath new deposit zone of province, notably on agro-techniques exhibited additional production, nutritional worth, and economic production of the system. This could even be additional encouraging with the mixing of aquatic crops in conjunction with fish cultures. The introduction of fish in conjunction with deep paddy in semi-deep to deep wetlands is common for the employment of food, energy, and total productivity in rainfed watershed coastal scheme of province. Likewise, integration of fish in conjunction with such aquatic starch and macromolecule made food crops, that is, water chestnut and makhana is the most significant farming system. Thus, the crop-fish diversity is far additional necessary and aimed substantial enough. Therefore, aquatic/wetland/lowland left a large array of scheme system services in Indo-Gangetic basin, if the correct water policy can be drawn at basin scale. It is therefore imperative to utilize this large unused scheme, scattered elsewhere, with impulsively for food, bread, and butter and economic stability that square measure inextricably coupled with the agricultural economy in these subzones as wider array of scheme and person.

In lowland ecosystem, farmers are habituated in each and every operational activity since time of immemorial. Not consistently, however, in scattered approach these marshy, fertile, productive aquatic systems square measure utilized sometimes by the associated farmers' for the assembly of nutritive and aquatic food crops, economically viable nonfood crops, medicinal, aquatic fodder crops, and different necessary aquatic plants together with biofertilizers. However, overall production of the system and also the economic production are not in the least encouraging thanks to technical copy and monetary stringency. In fact, these fertile and productive soils of soil will effectively be utilized following correct agro-techniques for higher production, which can flip economically viable for the agricultural sector.

13.10 CROPS SUITED UNDER THE ECOSYSTEM

A. Aquatic food crops

  i.   Deepwater rice (*Oryza sativa* L.)
  ii. Deepwater rice cum fish culture
  iii. Water chestnut or *singhara* (*Trapa bispinosa* Roxb.)

   iv. Makhana or gorgon nut or fox nut (*Euryale ferox* Salisb.)

   v. Taro or *panikachu* or *jalkachu* (*Colocasia esculenta*)

B. Aquatic food cum ornamental plants

   i. Lotus or padma (*Nelumbium speciosum*)

   ii. Water lily or saluk (*Nymphaea* spp.)

   iii. Royal water lily (*Victoria amazonica*)

C. Aquatic nonfood commercial crops

   i. Matreed (*Cyperus* spp.)

   ii. Shitalpati (*Clinogyne dichotoma*)

   iii. Hogla (*Typhae lephantina*)

   iv. Rattan palms or cane (*Calamus* spp.)

   v. Shola (*Aeschynomene aspera*)

D. Aquatic fodder crops

   i. Para grass or water grass (*Brachiaria mutica*)

   ii. *Coix* or gurgur grass (*Coix* spp.)

E. Aquatic medicinal plants

   i. Brahmi (*Bacopa monnieri*)

   ii. Kesuti or keshori (*Eclipta alba*)

   iii. *Ipomea* or swamp cabbage (*Ipomea aquatica*)

   iv. Talmakhana or kulekhara (*Hygrophila auriculata*)

   v. *Enhydra* or helencha (*Enhydra fluctuans*)

   vi. Indian pennywort or thankuni (*Hydroctyle asiatica*)

   vii. *Marsilea* or susni (*Marsilea quadrifolia*)

   viii. Water cress or swamp forest (*Nasturtium officinale*)

F. Aquatic aromatic plant

   i. Bach (*Acorus calamus*)

G. Aquatic weed cum plants of organic and biofertilizers

   i. Water hyacinth (*Eichhornia crassipes*)

   ii. Duck weed or taka pana (*Pistia stratiotes*)

iii. Jal jhangi or Hydrilla (*Hydrilla verticillata*)
iv. Azolla or pana (*Azolla pinata*)
v. Algae

13.11 IMPORTANCE OF WETLAND/LOWLAND ECOSYSTEM

a. Survival of human civilization is inextricably linked with these vast ecology.
b. Aquatic wetlands sustain economic stability to hundred millions of people.
c. On a short time scale, the ecosystem is useful as sources, sinks, and transformers of a multitude of chemical, biological, and genetic materials.
d. The system could be made effective to clean the polluted water, prevent flood and recharge groundwater aquifer, thereby increasing water productivity.
e. The valuation of aquatic ecosystem could add effectively as service provider with high contribution to improve water policy in Indo-Gangetic basin.
f. The system supports and sustains as a unique habitat for a wide variety of flora and fauna.
g. Rice, fish, other food and nonfood crop production is derived from the ecosystem.
h. A vast area of coastal wetland is very significant today for the conservation of mangrove forests as well as wildlife when environmental pollution is a burning question of the world.

13.12 IMPORTANCE OF WETLAND IN WATER PRODUCTIVITY

There is a greater scope for integrated farming in wetland, particularly for paddy cum fish culture or paddy-fish-vegetable farming in medium to low-lying situation of humid to subhumid areas. It is a very common practice for the cultivators of low-lying deepwater areas to utilize indigenous fish, found in paddy fields to supplement their rice diets. The relationship between rice and fish culture is obvious, but the technique and economic aspects require further study. Nowadays, the technique of fish cultivation in low-lying deepwater fields has received some attention and it seems that future prospects appear promising. The culture of fish along with deepwater paddy is likely to yield better return to the farmers as well as to the fishermen.

In areas where water retains for 3 to 8 months in the paddy fields, this system provides an additional supply of fish. It will, however, contribute substantially to increase the production of fish per hectare from inland water sources, as because, waterlogged paddy fields are the ideal natural habitats of various types of fish including indigenous ones.

Meeting the challenges of sustaining food security and economic growth which will require a lot, besides suitable rice production technologies development of improved farming practice for rainfed lowland environment that will diversify the farms through integration of agriculture, aquaculture, birds, animals, and other components. Such approach can ensure higher and stable farm productivity, income and employment opportunity without degrading the environment. Encouragement of scientific culture of integrated deep-water rice-fish, rice-fish-vegetables farming in waterlogged wetland areas is, therefore, most desirable due to the following sharing of advantages, which could be possible to utilize at its maximum levels. The synergistic effect of fish on rice yield; control of aquatic weeds and associated insects by fish; increased efficiency of resource utilization, reduced investment risk through crop diversification and additional sources of food and income; and more frequent visits to the field particularly for fish genotypes by the farmers, resulting in better crop management; low risk for poor deepwater rice farmers and modest capital investment; year round employment opportunity for the farm family and consequently, improvement of farm family income and nutrition level. Emphasizing the resource utilization, presently two major systems of integration of paddy, fish, and vegetables may be undertaken in the freshwater wetlands, are as follows:

Paddy-cum-fish culture and paddy-cum-fish-cum-vegetable culture and other suitable technologies to get highest return from the system.

13.13   WETLAND RESEARCH

   i.   To conduct and promote analysis on all types of interior aquatic ecosystems, particularly in under-explored areas and fewer understood topics.
   ii. To arrange a listing and assessment of aquatic diversity and its functions all told interior waters.
   iii. To ascertain a comprehensive ICT-enabled info of all types of data on all interior waters to function a web call support system.
   iv. To develop or adapt, and to contribute to the event or adaptation of, ways and applicable technologies for the restoration and management of interior aquatic ecosystems.

Wetlands square measure "lands shift between terrestrial and aquatic systems wherever the water table is sometimes at or close to the surface or the land is roofed by shallow water" and "must have one or additional of the subsequent three attributes: (1) a minimum of sporadically, the land supports preponderantly hydrophytes; (2) the substrate is preponderantly undrained hygrophytic soil; and (3) the substrate is nonsoil and is saturated with water or coated by shallow water at a while throughout the season of each year.

13.14   RAMSAR CONVENTION, 1971

Ramsar Convention focuses on areas of marsh, fen, vegetable matter land or water, whether or not natural or artificial, permanent or temporary, with water that is static or flowing, fresh, briny or salt, together with areas of marine water, the depth of that at low water does not exceed 6 m. Places where water is the first issue dominant plant and animal life and additionally the broader atmosphere, where the natural object is at or close to the land surface. Water is the most important substance for the existence of life on the planet. The highly uneven surface of the Earth has many areas where the land is saturated with or is submerged under the water that flows or remains standing for different duration from days to centuries before it returns to the oceans and the atmosphere. These watery areas were inhabited by a large diversity of plants and animals. The foundations of human civilization were laid on these lands. Fish were harvested long before the humans learned to grow food on the floodplain of Tigris and Euphrates rivers in the Middle East, and discovered rice and domesticated it in Eastern Asia. Many plants, that is, lotus and *Euryale ferox* and animals such as turtles, crocodiles, and swans became an integral part of the sociocultural ethos of the people in South Asia.

Jute was domesticated for fiber in India whereas in Egypt the giant sedge, *Cyperus papyrus*, led to the discovery of paper and was used to build boats that could be sailed across the ocean. In Europe, the reeds (*Phragmites australis*) were extensively used for thatching. In Asia, humans were not only attracted by the serene beauty of these watery areas that is reflected in their art, but also developed such compassion for the wildlife that many species were bestowed with divinity, associated with gods. Lotus became a symbol of sacredness and purity in both Hinduism and Buddhism. In India, these watery vegetated habitats were called Anup. Similar habitats elsewhere in the world were given many local names of which the most common in English were marsh, swamp, bog, and mire which differed greatly in their characteristics and biota. By the early 20th century, these habitats which hosted thousands of migratory birds from distant parts of the globe had turned into sport hunting grounds and soon concerns were raised in many countries for their protection. The term became internationally popular after the representatives of a few international organizations and national governments signed, on 2nd February, 1971 an agreement called as the Ramsar Convention. The Convention which originally emphasized the conservation and wise use of wetlands primarily as habitats for waterfowl has gradually extended its scope to larger array of aquatic ecosystems. Thus, wetlands have turned into waters without requiring the role of "land" in them.

13.15   DISTRIBUTION OF WETLANDS

Wetlands occur in all climatic zones from tropical deserts to cold tundra, and at all altitudes—from below the sea level to about 6000 m elevation in the Himalaya. Wetlands occur wherever water accumulates for enough long periods that allow the establishment of plants and animals adapted to the aquatic environment. Water need not be present permanently and the depth may generally fluctuate. Thus, wetlands occur in or along all water bodies—from temporary ponds to shallow or deep lakes, springs, streams, and rivers. Typically, wetlands are recognized by the presence of aquatic plants (called macrophytes) other than microscopic algae (phytoplankton or filamentous algae). The macrophytes play the most significant and predominant role in determining the functions of all wetlands. The growth and distribution of the macrophytes (other than the free-floating plants like duckweeds, Azolla, and water hyacinth) is determined among various factors by the water depth and is usually restricted to a depth of 2 m. Submerged plants may occur under clear water conditions to a depth of about 4 m. Therefore, only the shallow and usually the periodically flooded marginal areas of large rivers (called the floodplains) and lakes and reservoirs (called the littoral zones) are considered to be proper wetlands. A similar situation exists in the case of another kind of wetlands; the mangroves which also lie between higher land and the deep open waters of the sea.

Many wetlands have been modified and are managed by humans for specific purposes. For example, majority of the paddy fields and fish ponds have been created out of the natural wetlands by manipulating their vegetation and fauna. Similarly, innumerable tanks have been created in the arid and semiarid regions by blocking the runoff of small seasonal streams to store water for irrigation and or domestic uses. Thousands of large and deep reservoirs have been created by constructing dams over the rivers. The water level in these human-made water bodies fluctuates considerably and owing to the morphology of their basins, they have large littoral zones which have gradually turned into wetlands with rich aquatic vegetation and other biota.

Hundreds of thousands of human-made wetlands owe their existence to a wide range of human activities such as excavation of soil for making bricks or roads, stone quarrying, or due to water logging of low-lying lands along the canals. According to the latest estimate of wetlands in India, the human-made inland wetlands cover about 37% area (3,941,832 ha) and the remaining 63% are the natural wetlands (6,623,067 ha). There are also 4,140,116 ha of coastal wetlands (of which the intertidal mudflats of Kutch alone contribute about 51%) and 5,55,557 ha of wetlands smaller than 2.25 ha each. It is noteworthy that the paddy fields were included as wetlands in this inventory.

For centuries, we have looked at wetlands as forbidden, mysterious places: sources of pestilence, home to dangerous and pestiferous insects, and the abode of slimy, sinister creatures that rise out of swamp waters. They have been looked upon as places that should be drained for more productive uses by human standards: agricultural land, solid waste dumps, housing, industrial developments, and roads. One reason for the persistent under valuation of wetlands is that, historically, ideas of quantity are supported by an awfully slender definition of advantages. Economists have seen the worth of natural ecosystems solely in terms of the raw materials and physical product that they generate for human production and consumption specially specializing in industrial activities and profits (Dhindwal and Poonia, 1994). These direct uses but represent solely a little proportion of the whole price of wetlands that generate economic advantages so much in way over simply physical or marketed product. Staring at the whole quantity of a land basically involves considering its full vary of characteristics as an integrated system—its resource stocks or assets, flows of environmental services, and therefore the attributes of the scheme as a full. Broadly speaking, outline of the whole quantity of wetlands includes:

i. Wetland raw materials and physical products which are used directly for production, consumption and sale, such as those providing energy, shelter, foods, agricultural production, water supply, transport, and recreational facilities.

ii. The ecological functions which maintain and protect natural and human systems through services, such as maintenance of water quality and flow, flood control and storm protection, nutrient retention and microclimate stabilization and the production and the consumption activities they support.

The importance of wetlands has been realized throughout the world. The study of these ecosystems and scientific management still lags considerably behind the need.

Many researches on wetland and its impact on vegetation, animal, food chain, soil, waste management, etc. have done. The research works are as follows:

13.16   WETLANDS RESEARCH

Wetlands occupy approximately 6.4% of the world total area (Anonymous, 1986). The approximately area covered by wetland in the world, India and West Bengal are as follows:

The coast where land and sea intermingled with each other on which about 60% of the world population lives within 100 km of coastline make the lure of the coastal ecosystem. Being one of the most productive ecosystems, Indian coastlines spread over 7500 km with continental shelves of 0.50 km^2 and exclusive economic zone of 2.02 million km^2. Wetlands are very variable within biomass and one biome can contain a wide array of wetland types. Some of the properties and functions studied separately are common to all wetlands. Some work done in the past are on "the synergistic effects of hydrology, chemical inputs, and climatic conditions on wetland productivity and on how plants and animals adapt to stressful situations in various wetland types has provoked ideas for further research on wetlands where there is need of integration of several discipline." Wetlands are useful transformations of a multitude of chemical, biological, and genetic materials. They are sometimes described as the kidneys of the landscape for the functions they performed in hydrologic and chemical cycle. Thus, a most productive zone comprises between the dry, terrestrial, and permanently wet ecosystem, habitats of varieties flora and fauna. Wetlands are often the last portions of a landscape converted to alternative uses. Because, several wetlands are adjacent to surface waters, they typically represent the most effective chance for natural improvement of water quality as a result of their filtering and transformation capability. All wetlands, including those with high flows of water, tend to recycle nutrients repeatedly. A survey of the sea marginal wetlands along the Odisha coast shows that many of the wetlands present in this belt are dead or dying; Chilka lake itself is categorized under the "active but dying," the reason being natural geological phenomena or due to human activities. James (1995) also termed wetland as "Nature's Kidney."

In an urban environment, the wetlands may provide the only refuge for many kinds of wildlife, protect large amounts of valued property against flooding, serve as the main remaining mechanism for natural improvement of water quality, and recharge groundwater. Chaurs are areas that are inundated by rivers or rainwater from catchment areas that dry up in summer. This was observed by Rai and Munshi (1982). Agriculture in such areas is not very common because of the risk of flooding.

The aspects of benefits of wetlands are given with the following specific examples:

The understanding of the trophic structure of wetlands is very important for their management. A comprehensive account consists of the trophic structures of some typical wetlands in Kashmir. The study relates to five freshwater wetlands, viz. Nowgam, Mirgund, Malgam, Haigam, and Hokarsar. A total 190 species of phytoplankton were recorded from the wetlands. In all the wetlands, Chlorophyceae dominated 68–88% of the total biomass. Wetlands play an integral role within the ecology of the watershed. The mix of shallow water, high levels of nutrients, and first productivity is good for the event of organisms of that kind at the bottom of the food cycle and feed several species of fish, amphibians, shellfish, and insects. Several species of birds and mammals accept wetlands for food, water, and shelter, particularly throughout migration and breeding. The cultivation of freshwater and marine species is done in the wetlands. The protein yields from rice in the tropics, under good soil in wetlands and irrigation, are equaled or exceeded by the equivalent yield of coastal wetland fish. The lowland rice crop often faces moisture stress due to excess water that not only submerges the land but also sometimes the crop as well. The rice-fish integrated system in rainfed lowland rice fields is conceptualized by ultimately aiming at an environment that promotes synergism between the enterprises through recycling of wastes of one another leading to ecosystem conservation. The integrated farming in wetlands not only plays a significant role but also provides more opportunities to the farmers as well. However, it can overcome the risk to help secure higher income and employment opportunities (Pramanik and Mallick, 1996).

Phosphorus is one of the most important chemicals in the wetland ecosystems. It has been described as a major limiting nutrient in the bogs. Limitations of the elements have been noticed at least temporary for the short marshes. In water-saturated soil of wetlands, diffusion of oxygen into the soil is drastically reduced and in aqueous solution, it is estimated to be

1000 times slower than oxygen diffusion through a porous medium, such as drained soil. Nitrogen fixation, immobilization, and ammonia volatilization are the processes that are important in many wetlands. So, that nitrogen is often the most limiting nutrient in flooded soils, whether the flooded soils in natural wetlands or agricultural wetlands such as paddy field. Limitations of the elements have been noticed at least temporary for the freshwater inland marshes and freshwater tidal marshes. Fe and Mn in their reduced form reach toxic concentration in wetland soils. Ferrous ion diffusing to the surface of the roots of wetland plants may be oxidized by oxygen leaking by root cells in mobilizing phosphorus and coating roots with a Fe oxide causing barrier to nutrient uptake. Phosphorus is one of the most important chemical in the freshwater marshes. On deepwater swamps of Kerala, the soil are acid saline (ph 3.5–5.5) and are periodically subjected to inundation by seawater although big bunds strengthened by mangrove plants or bamboo or plaited coconut plants are used to block out tides. Pollen analytical studies and an examination of peat soil in wetlands claimed that the area comprising under the mangrove swamps of the Sundarbans was 5000 years back. Physical chemical and chemical characteristic of soil and water under rice-fish seed farming in rainfed lowland (0–50 cm water depth) were analyzed to prove the above findings.. The measurement of the atmosphere in the past decade taken all over the world revealed that concentration of methane had been rising at an annual rate of 1.0–1.3%, Carbon dioxide and N_2O at 0.4% and chlorofluorocarbon (CFC) rising at 5–6%. If the concentration of the gases having green house (warming) effect continues to rise at the present place, the average global temperature will increase by 1.5–4.5°C by 2030 AD and the level of water would arise by 56–345 cm by 2100 AD to flood the lowering coastal areas. Some estimates that methane emission from wetland rice fields in the world is 110 metric ton/year, which represents 21% of the total global emission of this gas.

## 13.17	RESEARCHES ON FISH CULTURE IN WETLANDS

Fishing is age old practice coming from generation to generation and having wide fish market worldwide. Inland fish culture requires wetlands, pond, lake, and river or fish tank. Both inland and marine fish culture requires wetland for their breeding and fish production. Fishing business and fish industry provides 2 million jobs in the country. The inland fresh-water fish culture in the wetland plays a significant role in the economic utilization of the waste wetland in the production of food as dual purpose.

It is thus possible for development of integrated aquaculture and agricultural system because in most of the states of eastern India, rice and fish both are the staple food of the majority of the population. The protein yields from rice in the tropics, under good soil of wetlands and irrigation are equaled or exceeded by the equivalent yield of coastal wetland fish. Shrimps and fish culture is the main economic sector in mangroves forests in areas where water retains for 3–8 months in the wetlands; the system provides an additional supply of fish. The fishes, which are good surrogates for aquatic biodiversity, are sensitive to alteration of habitat, including wetlands. India ranks a remarkable position in fish production in the world scenario and is the second largest producer of inland fishes both from culture and capture of fisheries. This has been possible with 2.25 million ha of ponds and tanks, 1.30 million ha of lakes, 2.09 million ha of reservoirs, and 1.23 million ha of brackish water areas, which are the potential of aquaculture in our country. About biodiversity, estuarine and freshwater fish are diverse and constitute more than 300 species of fishes, more than 250 species of plants, 150 species of birds, and innumerable other vertebrates and invertebrates. A method of consensus building for management of wetlands and fisheries using a systematic approach to participatory planning was initially developed in Bangladesh. Productivity of emergent vegetation in Indian Tropical Wetlands has been described by Rai and Datta (1982). The pneumatophores, prop roots, still roots, and the lower trunk regions of *Avicennia* sp. *Sonneratia* sp. and the members of Rhizophoraceae hold the dense cover of Bostrychileum like *Bostrychium* sp., *Caloglobassa* sp., *Centilla* sp. and *Murrayella* sp. These periphytons also have much value as fish food in the mangrove forests. The oxygen diffusion in the case of many wetland plants even takes place through the roots and oxidizes the adjacent anoxic soil. A total of 117 species of macrophytes are present in the aquatic and marshland habitat of Srinagar. Oxygen diffusion from roots in wetland condition is an important mechanism to overcome toxic effects of soluble reduced ions. Water conservation mechanisms have additional function in the wetland plants of reducing the rate at which soil toxins enter the plant through roots. This increases the probability of detoxifying them as they move through the oxidized atmosphere. Herbaceous wetlands are formed along the banks of rivers, lakes, reservoirs, and other freshwater bodies. Depending upon the water depth and its duration, its vegetation is characterized by the preponderance of (1) emergent, (2) rooted, (3) floating, and (4) submerged plant parts. The emergents have been further categorized into (a) tall growing (with the shoots emerging out of water more than 100 cm), (b) low growing (with the shoots height between 25 and 100 cm), and (c) ground layer species (with shoots less than 25 cm in height). In wetland, the species *Excoecaria agallocha* form almost pure vegetation and occupy 60–70% of the total area. In wetland, flooding stimulates ethylene production as well as cellular activity in the cortical cells of a number of plant species with the subsequent collapse and disintegration of cell walls. During high tides, large areas are inundated by brackish water, while low tides expose vast areas covered either by dense mangroves or halophytic herbs, shrubs and trees, flat river banks and sand dunes, without any forest formations in the surrounding areas of a wetland. Wetland plants have often been described as "Nutrient Pumps" that bring nutrients from the anaerobic sediments to the above ground strata. Phytoplankton in lakes and estuaries can be viewed as "Nutrient Dumps" that take nutrients out of the aerobic zone and through settling and death, deposits the nutrients in the anaerobic sediments. They have different functions in the nutrient cycling. The vegetation of the mangroves as well as other trees, climbers, shrubs, and grasses are found in the Sundarbans. These areas, in recent years, have turned out to be rich resources of lobsters, shrimps, fishes, and other sea animals. The varying vegetation provides shelter and food (algal flora, phytoplankton, and zooplankton, etc.) to the aquatic animals. The formation of vivipary among the mangroves species is very interesting as zygotes develop without any interruption through embryo and are able to produce seedling without intervention of any resting stage in wetland areas. The roots in many mangrove species like *Ceriops* sp., *Rhizophora* sp., *Bruguiera* sp., *Kandelia* sp., *Sonneratia* sp. and *Heritiera* sp. are negatively geotropic, that is, they come out of the soil in the form of vertical stalks called pneumatophores in wetland areas. Naskar and Bakshi (1987) observed that, in halophytes, the palisade tissues are massive, intercellular spaces are small and often absent, stomata either sunken or at near the level of epidermis, the other walls of epidermal cells cuticularized and thick. Naskar and Bakshi (1987) also observed that in the Sundarbans mangrove swamps, diatoms form a predominant group; the common ones identified from fishery water are *Navicula* sp., *Pleurosigma* sp., *Cymbella* sp., *Cyclotella* sp., etc. Naskar and Bakshi (1987) also studied about the common benthic flora of the brackish water of the fisheries of the Sundarbans that is *Oscillatoria* sp., *Gleocapsa* sp., *Symploca* sp., *Protococcus* sp., etc. Riverine wetland or moist land forests as typical edaphic types. These form distinct communities in tropical, subtropical, and temperate climates and are preserved in the several stages by their particular habitats. Freshwater swamp forests are usually found in permanently water-logged areas in the floodplains of rivers and usually of three subtypes: (1) Myristica Swamp Forests, (2) Tropical Hill Swamp Forests, and (3) Creeper Swamp Forests. Some typical plants are found in the marshy lands of Manipur.

13.18 RESEARCHES ON IMPACT OF WETLANDS ON ANIMALS

Most of the North American bird species prefer wetlands. Indian wetlands have been destroyed, bird populations have slowly decreased. The continental duck breeding population declines from 45 million to 31 million birds, a decline of 31%. During 1980, 75% of forest dwelling species are geotropically migrants, many of whom rely on coastal wetland habitats during their arduous migrations. About six major kinds of adaptation by aquatic or amphibian organisms are usually observed to control gaseous exchange: (1) development and modification of specialized regions of the body for gaseous exchange, (2) mechanism to improve the oxygen gradient across diffused membrane, (3) internal structural changes, (4) modification of respiratory pigments to improve oxygen-carrying capacity, (5) behavioral patterns, such as decreased locomotors activity or the closing of a shell during low oxygen stress, (6) physiological adaptations. The importance of river bottomlands to wildlife relative to adjacent uplands in Illinois. The enzymes of salt-tolerant yeasts seemed to be salt-sensitive and it has been suggested that the organic compounds act as "Compatible Solutes" that raise the osmotic pressure without interfering with enzymatic activity. Wetlands perform some functions, such as maintenance of breeding habitat for some bird species that are either unique or particularly efficient in proportion to their size. River floodplain wetlands form natural corridors for the migration of fish, birds, mammals, and reptiles. The availability of large riparian areas, which include wetlands, is the primary factor that explains the number of birds that breed at high elevations in Central Arizona. Wetlands as islands are a terrestrial sea and suggested that bird diversity follows the rules of island biogeography (more species with larger island area) as shown for Prairie potholes. On waterfowl, which provide some of the best long-term records of species that depend on wetlands, show steady declines. Many waterfowl species are sensitive to reductions in area, patch size, wetland density, and proximity to other wetlands.

13.19   RESEARCHES ON IMPACT OF WETLANDS ON NONFOOD MATERIALS

The genus Typha is represented by four species: *T. angustata* Bory et Chaub, *T. elephantiana* Roxb. *T. latifolia* L. and *T. laxmanii* Lepech. About 7500 lux is necessary for the development of rhizome systems in wetlands. Continuous growth under deepwater (50 cm or above) or sudden large fluctuations in water level affected the growth of *Typhaangustata* adversely. About the suckers or slips with rhizomes or subterranean, nuts are used as planting material in wetlands. For 1 ha, 15 quintals of planting materials (slips) are required in wetlands. The *Clinogyne* sp. provides pith which is usually discarded by mat makers but which can be used in paper manufacture.

13.20   RESEARCHES ON WETLANDS USED IN WASTEWATER MANAGEMENT

A very detailed work on freshwater wetland using for wastewater management is carried out by the Texas Department of Health. A book on "Water Hyacinth Culture for Waste Water Treatment" provides very useful information. The potential for using freshwater wetland for wastewater management has received much attention in North America during recent years.

13.21   RESEARCHES ON WETLAND AS SUSTAINABLE RESOURCES

Wetland management plays a key role to protect, restore, manipulate, and provide for functions. Wetlands provide many important services to human society, but are at the same time ecologically sensitive systems. For sustainable management strategies for wetlands, natural and social sciences can contribute to best output. Sustainable resource is used as one with long-term economic viability besides having both ecological and social compatibility. The decisions about wetlands' future use might be accompanied by significant resource saving as a result of the development of rules of thumb, linking the characteristics of wetland vegetation with the functions, which wetlands perform. Modeling the factors determining ecological and economic value gives rise to conceptual and methodological problems. Ecologists, however, are more concerned with ecological values, which provide an underlying long-run notion of value interpreted in a more general sense. This chapter investigates the nature of the link between these two aspects of value in the context of a wetland in Northern India, which is also designated as a Ramsar site and a national park.

13.22 CONCLUSION

Wetlands directly or indirectly have an enormous ecological, biological, economic, commercial, and socioeconomic importance and values. Such lands contain very rich components of the biodiversity profuse life flora and fauna of important local natural and regional significance. Practically, farmers are using these vast wetlands as per their traditional need, which are not always agro-ecologically suitable, subject to vacant mostly without any proper use. The valuable natural resources of wetlands should be used to realize their fullest economic return and at the same time retaining their ecological and environmental integrity. Modern technology of watershed management and culture of different types of animals and their integration with crop production, even in small farms, will not only replace the depletions but also increase the production and would affect the sustainable development of the wetlands with renewable locally available inputs. The venture will be economically profitable, ecologically viable, socially acceptable as well as very efficient in the utilization of available resources.

Assessment of the full values of wetlands to communities and to society as a whole, and the real costs, social as well as economic, of their destruction or improper/inappropriate utilization provides the basis for competing more effectively with assessments based on alternative economic uses.

The main objective of wetland conservation should be that they are protected and managed in order to maintain the optimum range of sustainable benefits for mankind and wildlife for environmental quality.

i. Technologies for utilization of agricultural residues to the field: Return of agricultural residues to the field can supplement, renew, and accumulate organic matter in soil and sustain the productivity.

ii. Application of manure resulting from manure pits and biogas generation tanks to the field: For effective utilization of manures, animal sheds can be put up near the wetlands to reutilize their excreta for feeding fish and encouraging growth of planktons in the tank beds and then again as and when needed using the tank sludge and weeds for applying them to the field.

iii. Breeding fish in paddy fields, particularly in lowlands and deepwater rice, for prevention and control of plant diseases, elimination of pests and weeds.

iv. Control of insect pests through the use of fishes, birds, and frogs: Indiscriminate capture of fishes and frogs and excessive hunting of birds are harmful and these should be avoided. This is very important as they control the pest population and maintain the balance of ecosystem.

v. Increasing the stability of the production system through crop/livestock/soil interaction over time: Sustainable land use system with comprehensive management and coordinated development of crop production, mulberry production, animal production, and fishery can be possible in a better way in wetland situations than in other environment.

vi. Planting trees and grasses for conservation of water and soil. This is particularly important when the soil surface is undulated.

The sustainable utilization of wetlands regulates the economic welfare of the people to a considerable extent as because a large sector of the human population depends on wetlands for their everyday survival. The initial effort of utilizing the wetland ecosystem can be considered as a model, on the basis of which further modifications might be incorporated to derive the most gainful combination. Importantly, the involvement of the farmers of the area through farmer's cooperative society could be very important in this effort to transform potentially fertile but circumstantially low productive region to make it highly remunerative with increased farm income, more employment as well as providing better nutrition for the consumers.

KEYWORDS

- **inland wetland**
- **coastal wetland**
- **groundwater recharge**
- **conservation and management**

CHAPTER 14

Water Pollution in Agriculture

ABSTRACT

Agricultural pollution refers to biotic and abiotic by-products of farming practices that result in contamination or degradation of the environment and surrounding ecosystems, and/or cause injury to humans and their economic interests. The pollution may come from a variety of sources, ranging from source pollution (from a single discharge point) to more diffuse, landscape-level causes, also known as nonpoint source pollution. Management practices play a crucial role in the amount and impact of these pollutants. Management techniques range from animal management and housing to the spread of pesticides and fertilizers in global agricultural practices. These have got relevancy with water resources and its utilization.

14.1 INTRODUCTION

Water is a precious commodity for healthy living. Industrial growth, urbanization, and therefore the increasing use of artificial organic substances have serious and adverse impacts on water bodies. Polluted water contains chemicals which result in health hazards and water-borne diseases. The groundwater and surface water are currently contaminated with serious metals; pesticides residue creates health problems. Water access may be outlined because the range of individuals use safe water and comfortable drinking water. There should be a trial to sustain it, and there should be a good and equal distribution of water to all or any segments of the society. Urban areas usually have the next coverage of safe water than the agricultural areas. Even at intervals of a section there is variation.

14.2 WATER CONTAMINATION IN URBAN AND RURAL AREAS

a. Pesticides

Residual effects come through runoff from agricultural lands and contaminate all types of water. Leachate from lowland sites is another major contaminating supply and its effects on the ecosystems are adverse.

b. Sewage

Untreated or inadequately treated municipal biodegradable waste may be a major supply of groundwater and surface pollution within developing countries. The organic material is discharged with municipal waste into the water and uses substantial chemical element for biological degradation of the ecological balance of rivers. Biodegradable pollution conjointly carries microorganism and causes harmful effect in human health.

c. Nutrients

Polluted water comes from household, agricultural field, and industrial effluents and contains phosphorus and element, chemical runoff, manure, increase the extent of nutrients in water bodies, and may cause eutrophication within the lakes and rivers. The nitrates return chiefly from the chemical that is additional. High fertilizers use causes nitrate contamination in groundwater, with the result that nitrate levels in drinking water is high than the critical levels. Sustainable agricultural practices will facilitate in reducing the number of nitrates within the soil.

d. Synthetic organics

Artificial compounds in use nowadays are found within the aquatic setting and accumulate within the organic phenomenon. Persistent organic pollutants represent the foremost harmful component for the scheme and for human health, that is, agricultural chemicals and pesticides. These chemicals accumulate in fish and cause serious harm to human health. Agricultural pesticides when used on a large-scale, the groundwater gets contaminated.

e. Acidification

During rainy season, runoff water comes from distance. This runoff water carries all the surface water containing organic and inorganic materials which is stored in river, lakes, and sometimes reservoirs or watershed. This water contains industrial and agricultural waste material which is acidic in nature.

14.3 DRINKING WATER POLLUTION

Water pollution can be either naturally or anthropogenic and these pollutants are discussed below:

a. Fluoride
Water contaminated with fluoride is harmful for human bone; it causes bone deformities and dental caries. Mainly, its bad impact is on weakening of the bones, but higher concentration causes adverse effect on health. High fluoride is found naturally in Rajasthan, India.

b. Arsenic
Arsenic is a naturally occurring element particularly found on both banks of river Ganges and also on overexploitation aquifers and arsenate pesticides. High concentrations of arsenic in water and in food adversely affect the health. The high concentration of arsenic is found in drinking water as well as in all food stuffs in 12 districts of West Bengal. The major symptoms found in skin specially hands, palm, foot, and if it is severe, then it will also be found in the whole body as well as inside the body. Skin lesions can be easily found due to arsenic contamination in the groundwater and also due to drinking the same undergroundwater. An alternative drinking water source has been provided by the government side and different technology has come up to remove arsenic from water.

c. Lead
Lead contamination is common and everywhere we use such as pipes, household tools, instruments, and the service connections of some household plumbing systems also contaminate the drinking water with high level.

d. Recreational use of water
For everyday life, with known and unknown situation, we use untreated sewage, sludge, industrial effluents, and agricultural waste in different ways. Urban waste releases into water bodies, such as the ponds, tanks, lakes, coastal areas, and rivers endangering their use for recreational purposes, such as swimming and bathing and sometimes drinking purposes.

e. Petrochemical effluents
Petrochemical effluents contaminate the groundwater as well as surface water from underground petroleum storage tanks. Moreover, it also affects the agricultural land, human habitats, and recreational places.

f.　Heavy metals waste

Most of the heavy metals come from the site of waste pit mining waste and tailings, landfills, waste dumps, and urban waste pit.

g.　Chlorinated pollutants

In many industries, contamination of chlorine comes from the metal and plastic effluents, fabric cleaning, electronic, and aircraft manufacturing waste, and they contaminate the groundwater.

Drinking water causes more human illness than any other environmental impact. The diseases so transmitted are chiefly due to microorganisms and parasites. Among the various water-borne diseases, the notable example is cholera. Many harmful germs are spread from human excreta as well as from other animals. Most human pathogens can be classified as viruses, protozoa, helminthes, or bacteria. Both raw sanitary sewage and land runoff contain pathogenic organisms. A good number of enteric viruses found in water happen to infect various organs of the human body. The body burden of these toxicants increases day by day.

Therefore, it is essential to maintain water quality and prevent the pollution of both surface water and groundwater. The impact of arsenic present in the groundwater and its effect on the food chain with special reference to a small mouza, Ghetugachi, situated in Chakdaha Block of Nadia district were studied in detail. Groundwater extraction had started here extensively in the early 70s as a result of green revolution to encounter water demand for groundwater, and about 12 districts of West Bengal were severely affected. Arsenic contamination normally occurs within the depth range of 20–100 m, while more than 100 m deep tube wells are considered relatively safe, though questions are arising regarding the safety of the water drawn from these deep tube wells also. It is observed that shallow wells (less than 20 m) also tend to have low arsenic contamination. Arsenic content in groundwater with concentration ranges from 50–500 mg/l. The contamination is mostly found in upper deltaic plain and comprises of younger Ganga sediment. The organic content of the sediment is nearly 1%. The groundwater chemistry of the affected area reflects high concentration of bicarbonate, phosphate. nitrate, sulfate, and fluoride concentration is low. Field parameters show that pH is usually near neutral.

14.4　WATER POLLUTION IN AGRICULTURE

Crop yields in agricultural systems trusted internal resources, utilization of organic matter, constitutional biological management, and downfall patterns.

Production was safeguarded by growing quite one crop or selection in house and time in an exceedingly field as insurance against cuss outbreaks or severe weather. In return, crop rotations suppressed insects, weeds, and diseases by breaking the life cycles of those pests and diseases. A typical maize farmer grew rotating maize with many crops as well as soybeans, and tiny grain production was intrinsic to take care of farm animal. Occasional employee and no specialized instrumentation accomplish the different types of field work. During this form of farming systems, the link between agriculture and ecology is degrading. In fact, many agricultural scientists have got hold of a general accord that trendy agriculture confronts an environmental crisis (Altieri, 1995).

Growing variety of individuals became involved regarding the long property of existing food production systems. Proof has accumulated showing that; whereas the current capital-oriented and technology-intensive farming systems are extraordinarily productive and competitive. They conjointly bring a spread of economic, environmental, and social issues (Castaneda and Bhuiyan, 1988).

Nowadays additional and additional farmer's square measure is rewarded by economies of scale. Consequently, lack of rotations and diversification subtract key self-activating mechanisms, turning monocultures into extremely vulnerable agro-ecosystems obsessed with high chemical inputs. The pollution might return from a spread of sources, starting from beginning pollution, landscape level causes, conjointly called nonpoint supply pollution. Management practices play an important role within the quantity and impact of those pollutants. Agronomic management techniques vary from animal, pesticides, and fertilizers.

a. Pesticides

Pesticides and herbicides square measure applied to agricultural land to regulate pests that disrupt crop production. Pesticides retention in soil may alter microorganism processes, may increase plant uptake of the chemical, and conjointly cause toxicity to soil organisms. The extent to which the pesticides and herbicides persist depends on the compound's distinctive chemistry, which affects natural process dynamics and ensuing fate and transport within the soil surroundings. Pesticide's natural action happens once pesticides combine. The quantity of natural action is related to with specific soil and chemical characteristics, and therefore the degree of precipitation and irrigation. Natural action can possibly happen if a soluble chemical is employed; high watering happens simply once chemical application is done.

Modernization in agriculture has progressed. The ecology-farming linkage was often broken as ecological principles were ignored and/or overridden. The modern agriculture confronts an environmental crisis and has become concerned about the long-term sustainability of existing food production systems. Evidence has accumulated showing that whereas the present is capital-oriented.

The nature of the agricultural structure and prevailing policies have led to this environmental crisis by favoring large farm size, specialized production, crop monocultures, and mechanization. Integration of agriculture components into international economies, imperatives to diversity disappear, and monocultures are rewarded by economies of scale. Crop diversification takes away key self-regulating mechanisms, turning monocultures into highly vulnerable agro-ecosystems. Pollutants from agriculture sectors consist in adoption of modern agro-technologies in agriculture. Harmful agents of agriculture field origins affect the environment.

b. Fertilizers

Leaching, runoff, and eutrophication play an important role for betterment of nitrogen (N) and phosphorus (P) applied to agricultural land will offer valuable plant nutrients. However, if not managed properly, excess N and P will have negative environmental consequences. Excess N equipped by each artificial fertilizers (as extremely soluble nitrate) and organic sources like manures result in groundwater contamination of nitrate. Nitrate-contaminated potable will cause infant syndrome.

c. Organic Contaminants

Manures and biosolids contain several nutrients consumed by animals and humans within the type of food. The returning of such waste product to agricultural land and manures and biosolids contain not solely nutrients like carbon, nitrogen, and phosphorus; however, they will additionally contain contaminants, as well as prescription drugs and private care product. There a large selection and immense amount is consumed by each human and animals, and every distinctive chemistry is present in terrestrial and aquatic environments.

d. Heavy Metals

Many heavy metal elements interfere with soil and water in agricultural system and it comes indirectly from farm to fork. These elements are arsenic, lead, fluoride, and mercury. Some farming techniques like irrigation, will result in the accumulation of element (Se) that happens naturally within the soil. This can end in downstream water reservoirs containing concentrations of element in that area unit which is deadly to life, livestock, and humans.

e. Land Management
Tillage and inhalation general anesthetic emissions: Natural soil biogeochemical processes end in the emission of assorted greenhouse gases, as well as inhalation general anesthetic. Agricultural management practices will have an effect on emission levels. Tillage levels have additionally been shown to have an effect on inhalation of general anesthetic emissions.

14.5 SOIL EROSION AND GEOLOGICAL PHENOMENON

Agriculture contributes greatly to eating away and sediment deposition through intensive management or inefficient land cowl. It is calculatable that agricultural land degradation is resulting in associate irreversible decline in fertility on regarding 6 million angular distance of fertile land annually. The accumulation of sediments in runoff water affects water quality in numerous ways in which geological phenomenon will decrease the transport capability of ditches, streams, rivers, and navigation channels. It may also limit the number of sunshine penetrating the water, which affects aquatic collection. The ensuing muddiness from geological phenomenon will interfere with feeding habits of fishes, touching population dynamics. Geological phenomenon additionally affects the transport and accumulation of pollutants as well as phosphorus and numerous pesticides.

14.6 BIOTIC SOURCES

Greenhouse gases from soiled waste: The globe organization, food and agricultural organization expected that 18 of phylogenes are greenhouse gases and they return directly or indirectly from the world's eutherian. This report in addition steered that the emissions from eutherian were larger than that of the transportation sector. Whereas eutherian will presently play employment in producing gas emissions, the estimates area unit is argued to be a deception. The world organization agency used a life cycle assessment of animal agriculture; they did not apply an identical assessment for the transportation sector.

14.7   BIOPESTICIDES

Biopesticides area unit pesticides are derived from natural materials, such as animals, plants, microorganisms, bound minerals. Some issues exist that biopesticides could have negative impacts on populations of nontarget species. Biopesticides area unit is regulated. As a result of biopesticides, area unit is less harmful and have fewer environmental effects than alternative pesticides.

14.8   SOCIAL AND ECONOMIC IMPLICATIONS OF POLLUTION

The specialization of production units has semiconductor diode to the image that agriculture may be a fashionable miracle of food production. Most agriculturalists had assumed that the agrosystem/natural ecosystem duality need not result in undesirable consequences. Diseases of the ecotype, embody erosion, loss of soil fertility, depletion of nutrient reserves, Stalinization and alkalization, loss of fertile croplands to urban development, and diseases. Agricultural systems, the number of energy endowed to because of pests in several crops, despite the substantial increase within the use of pesticides may be a symptom of the environmental crisis moving agriculture. Agricultural practices negatively have an effect on tormenter natural enemies and pests.

Fertilizers, on the other hand, are praised as being extremely related to the temporary increase in food production determined in several countries. National average rates of nitrate applied to most productive lands fluctuate between 120–550 metric weight unit N/ha. However, the bountiful harvests created a minimum of partly through the utilization of chemical fertilizers, have associated, and sometimes hidden, costs. A primary reason why chemical fertilizers contaminate the setting is because of wasteful application and also the undeniable fact that crops use them inefficiently.

14.9   CONCLUSION

Modern agricultural structure and contemporary policies have decidedly influenced the context of agricultural technology and production, which in turn has led to environmental problems of a first and second order. In fact, given the realities of the dominant economic milieu, policies discourage resource-conserving practices and in many cases, such practices are not privately profitable for farmers. Set of policy priority changes could be implemented for a renaissance of diversified or small-scale farms may be unrealistic, because it negates the existence of scale in agriculture and ignores the political state power set forth for globalization. A more radical transformation of agriculture is needed; one guided by the notion that ecological change in agriculture, a comparable changes in the social, political, cultural and economic arenas that also confirm agriculture. In other words, change toward a more socially just, economically viable, and environmentally sound agriculture should be the result of social movements in the rural sector in alliance with the urban organizations.

KEYWORDS

- **pesticides**
- **fertilizer application**
- **sewage**
- **synthetic organics**
- **petrochemicals**

References

Akanda, M. A.; Loof, R.; Islam, M. S. Irrigation Water Quality of Thabua Irrigation Project of Thailand. *Bangladesh J Agric. Res.* **2001**, *26*(4), 497–514.

Altieri, M. A. Agroecology. In *The Science of Sustainable Agriculture*; West View Press: Boulde; 1995.

Amarasinghe, U.; Shah, T.; Turral, H.; Anand, B. K. *India's Water Future to 2025–2050: Business-as-Usual Scenario and Deviations*. Colombo: International Water Management Institute IWMI Research Report 123, 2007; pp 41.

Anonymous. Indian Agriculture in Brief, 21st ed.; Ministry of Agriculture, Government of India, 1986.

Boland, J. J.; Whittington, D. The Political Economy of Water Tariff Design in Developing Countries: Increasing Block Tariffs versus Uniform Price with Rebate. In *The Political Economy of Water Pricing Reforms*; Dinar, A. Eds.; Oxford University Press and the World Bank: Washington D.C., 2000.

Bruns B. R.; Meinzen-Dick, R. Frameworks for Water Rights: An Overview of Institutional Options. In *Water Rights Reform: Lessons from Institutional Design, eds. Bryan Rudalph Burns, Claudia Ringer and Ruth Meinzen-Dick*; International Food Policy Research Institute: Washington, D.C., 2005.

Castaneda, A. R.; Bhuiyan, S. I. Industrial Pollution of Irrigation Water and its Effect on Rice Productivity. *Philipp. J Crop Sci.* **1988**, *13*(1), 27–35.

Das, P. Private Water, Public Misery: Flawed Policy, Frontline: April 21, 2006.

DFID (Department for International Development). Willing to Pay but Unwilling to Charge: Do 'willingness-to-pay' Studies Make a Difference? Water and Sanitation Program-South Asia, Department for International Development, June 1999.

Dhindwal, A. S.; Poonia, S. R. On-Farm Water Management a Success Story. *Intensive Agric.* **1994**, *32*(1–2), 37–38.

Falkenmark, M.; Lundquvist, J.; Widstrand, C. Macro-Scale Water Scarcity Requires Micro-Scale Approaches: Aspect of Vulnerability in Semi-Arid Development. *Nat. Res. Forum.* **1989**, *13*(4), 58–267.

Iyer, R. R. Water: Perspectives, Issues, Concerns; Sage Publications: New Delhi, 2003.

James, E. J. Managing the Wetlands and their Watersheds. *Yojana* **1995**, *39*(1 & 2), 43–50.

Johansson, R. C. Pricing Irrigation Water: A Literature Survey. Policy Research Working Paper 2449, Washington D.C: The World Bank, 2000.

Johansson, R. J. Micro and Macro-Level Approaches for Assessing the Value of Irrigation Water. Policy Research Working Paper 3778. The World Bank, 2005.

Kumar, D.; Singh, M. K.; Singh, O. P.; Shiyani, R. L. A Study of Water-Saving and Energy-Saving Micro Irrigation Technologies in Gujarat. Research Report 2. Anand: India Natural Resource Economics and Management Foundation, 2004.

Narayanamoorthy, A. Economics of Drip Irrigation in Sugar Cane Cultivation: An Empirical Analysis. In *Managing Water Resources: Policies, Institutions, and Technologies*; Ratna Reddy, V., Mahendra Dev, S., Eds.; New Delhi: Oxford University Press, 2006.

Naskar, K. R.; Guha Bakshi, D. N. *Mangrove Swamps of the Sundarbans - An Ecological Perspectives*; Naya Prakashan Pub.: Calcutta, 1987; pp 1–263.

NWDA (National Water Development Agency). Annual Report, 2012.

Onken, B. M.; Hossner, L. R. Plant Uptake and Determination of Arsenic Species in Soil Solution Under Flooded Conditions. *J Environ. Qual.* **1995,** *24,* 373–381.

Pramanik, M.; Mallick, S. Farmer's Participatory Approach for Improvement of Present Status of Irrigation Water Utilisation in DVC Canal Command. Proc. of ICID/FAO Workshop, Rome, Water Report 8, 1996.

Rai, D. N; Datta Munshi, J. S. In *Wetlands Ecology and Management*; Gopal, B., et. al., Eds.; Part-II, 1982; pp 89–95.

Rath, N. Linking Rivers: Some Elementary Arithmatic'. *Econ. Political Wkly.* **2003,** *38*(29), 3033.

Shah, T.; Verma, S. *Co-Management of Electricity and Groundwater: An Assessment of Gujarat's Jyotirgram Scheme*, Vol. 43; 2008.

Skogerboe, G. V.; Hyatt, M. L.; England, J. D.; Johnson, J. Raymond, Design and Calibration of Submerged Open Channel Flow Measurement Structures: Part 2 Parshall Flumes, *Reports,* 1967, p 81.

Smakhtin, V.; Anputhas, M. An Assessment of Environmental Flow Requirements of Indian River Basins; International Water Management Institute, IWMI Research Report-107: Colombo, 2006, p 36.

Smakhtin, V.; Arunachalam, M.; Behera, S.; Chatterjee, A.; Das, S.; Gautam, P.; Joshi, G. D.; Sivaramakrishnan, K. G.; Unni, K. S. Developing Procedures for Assessment of Ecological Status of Indian River Basins in the Context of Environmental Water Requirements. International Water Management Institute, IWMI Research Report 114: Colombo, 2007, p 34.

Thakur, D. P; Thakur, M. K. *Integrated Agriculture Aquaculture- Animal Husbandry Farming System*, 1991, pp 475–484.

Thanh, N. C.; Biswas, A. K. *Environmentally Sound Water Management*; Oxford University Press: Delhi, 1990.

Trist, E. The Evaluation of Socio-Technical System: A Conceptual Framework and Action Research Programme, Toronto, 1980.

Zaman, A. Integrated Water Resources Management for Agricultural Sustainability. *Proc. International Conference on Environment and Development* Held at Science City, Kolkata on 4–8 February, 2004.

Zaman, A. Farmer-Managed Small-Scale Irrigation Systems for Sustainable Crop Production in Uganda, East Africa. *J Food Agric. Environ.* **2003,** *1*(3–4), 312–315.

Zaman, A.; Das, S. K.; Siddique, A. Inter-Basin Transfer of Water Through Interlinking Rivers for Harnessing Water Resources. Proc. International Conference on Regional Cooperation on Transboundary Rivers: Impact of the Indian River-linking Project (IRLP) Held on 17–19 December 2004 in Dhaka, 2004; pp 16–26.

Zaman, A.; Talukder, M. L.; Zaman, P.; Hedayetullah, M. Adoption of Improved Water Management Technologies for Sustaining Agricultural Productivity. *Int. J. Agron. Crop Sci.* **2016,** *1*(1), 1–7.

Zimmer, D.; Renault, D. Virtual Water in Food Production and Global Trade: Review of Methodological Issues and Preliminary Results. Proceedings of the International Expert Meeting on Virtual Water Trade, Value of Water-Research Rapport Series, 2003, 12, 93–109.

Index

For Product Safety Concerns and Information please contact our EU
representative GPSR@taylorandfrancis.com
Taylor & Francis Verlag GmbH, Kaufingerstraße 24, 80331 München, Germany